HAUNTINGS

www.penguin.co.uk

For more information on Neil Oliver and his books,
see his website at www.neiloliver.com

HAUNTINGS

A BOOK OF GHOSTS AND WHERE TO FIND THEM

NEIL OLIVER

bantam

TRANSWORLD PUBLISHERS
Penguin Random House, One Embassy Gardens,
8 Viaduct Gardens, London SW11 7BW
www.penguin.co.uk

Transworld is part of the Penguin Random House group of companies
whose addresses can be found at global.penguinrandomhouse.com

First published in Great Britain in 2023 by Bantam
an imprint of Transworld Publishers

A CIP catalogue record for this book
is available from the British Library.

ISBN 9781787636347

Typeset in 12.25/17pt Minion Pro by Jouve (UK), Milton Keynes
Printed and bound in Great Britain by Clays Ltd, Elcograf S.p.A.

The authorized representative in the EEA is Penguin Random House Ireland,
Morrison Chambers, 32 Nassau Street, Dublin D02 YH68.

Penguin Random House is committed to a sustainable
future for our business, our readers and our planet. This book is
made from Forest Stewardship Council® certified paper.

For Mum and Dad,
Norma and Pat

CONTENTS

A Map of British Hauntings   x–xi

Introduction   1

1.   Sandwood Bay, near Cape Wrath   9
2.   Aughrim Battlefield, County Galway   18
3.   Ladybower Reservoir and the Dambusters, Derbyshire   31
4.   Old Vicarage, Grantchester, and Rupert Brooke   43
5.   Pendle Hill and Witches, Lancashire   52
6.   Ben MacDhui and the Big Grey Man   63
7.   Iona and a Ghost of the Past   77
8.   Windsor Castle, Henry VIII and Herne the Hunter   89
9.   Glamis Castle: the Grey Lady, the Monster and the Tongueless Woman   102
10.   Raynham Hall and the Brown Lady   112
11.   Culloden Moor, the Two Sights and Dreams That Will Not Die   127
12.   The Tower of London, St Peter in Chains and Anne Boleyn   140
13.   Mary King's Close, the Wizard of West Bow and Little Annie's Doll   151

14. Borley Rectory, the Most Haunted House in England   163

15. Wistman's Wood, Devon, and the Deepest Roots   176

16. Wayland's Smithy   189

17. The Angel of Mons   200

18. Lud's Church and a Lollard Martyr Called Alice   213

19. St Peter's Church, Luddites and the Ghost in the Machine   224

20. Ernest Shackleton, *South* and the Fourth Man   239

21. Coffin Roads   253

22. The Skye Ferry and the Wee Black Car   266

23. Glencoe, and Wrongs Unforgotten, Unforgiven   277

24. Tigguo Cobauc, and the Fear and Lure of the Dark   291

25. Number 50 Berkeley Square   301

26. Ravenser Odd and For Whom the Bell Tolls   315

27. Mount Cottage, Dorset, and a Living Ghost   321

28. Fathers and Sons   336

Acknowledgements   343

Picture Acknowledgements   345

Index   347

'you want to know
whether i believe in ghosts
of course i do not believe in them
if you had known
as many of them as i have
you would not
believe in them either.'

DON MARQUIS, *ARCHY AND MEHITABEL*

A MAP OF BRITISH HAUNTINGS

Numbers refer to corresponding chapters:

● = Primary locations   ○ = Secondary locations

1 Sandwood Bay, near Cape Wrath
2 Aughrim Battlefield, Galway
3 Ladybower and Derwent reservoirs, Derbyshire
4 Old Vicarage, Grantchester
5 Pendle Hill, near Clitheroe, Burnley and Colne
5b Palace of Holyroodhouse, Edinburgh
5c Gallows Hill, Lancaster
6 Ben MacDhui, Cairngorms National Park
7 Isle of Iona
8 Windsor Castle
9 Glamis Castle, Forfar
9b Castle Hill, Edinburgh
10 Raynham Hall, Fakenham
11 Culloden
12 Tower of London
13 Mary King's Close, Edinburgh
14 Borley Rectory, Borley

21a Island of Scarp
21b Harris Coffin Road
21c Mardale Green, Westmorland
21d Swaledale Corpse Way, linking Keld with Grinton
22 Skye Ferry
22b Barnhill, Jura
23 Glencoe
24 Mortimer's Hole, Nottingham Castle
25 50 Berkeley Square, London
25b Victoria Terrace, Dumfries
26 Ravenser Odd, Humber Estuary
26b Skara Brae, Orkney
26c Cardigan Bay
26d Dunwich, Suffolk
27 Mount Cottage, Dorset
27b Whitby Abbey, North Yorkshire
27c Blickling Hall, Norfolk
27d Preston Manor, Brighton
27e Newton House, Llandeilo

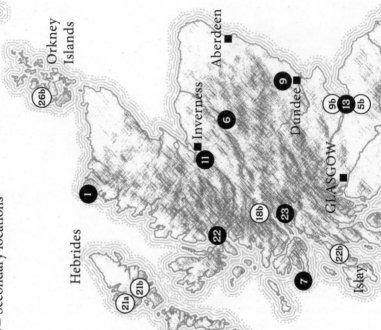

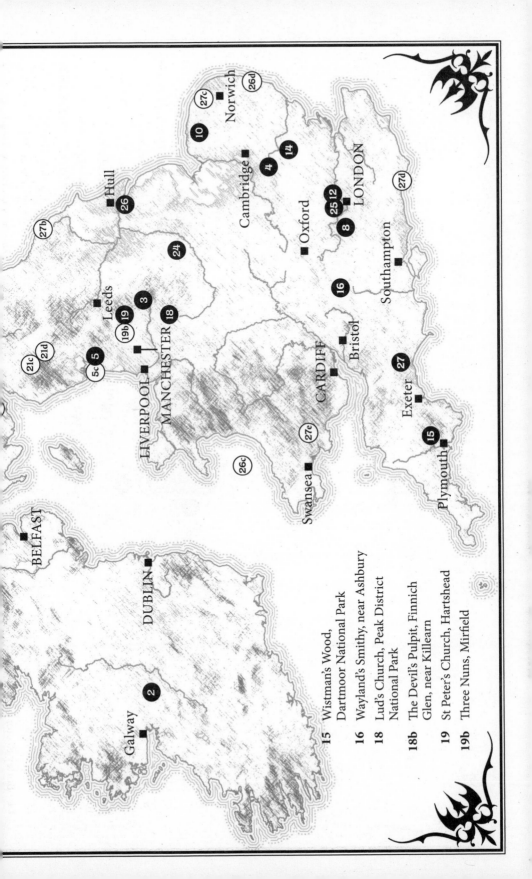

15 Wistman's Wood,
Dartmoor National Park

16 Wayland's Smithy, near Ashbury

18 Lud's Church, Peak District
National Park

18b The Devil's Pulpit, Finnich
Glen, near Killearn

19 St Peter's Church, Hartshead

19b Three Nuns, Mirfield

INTRODUCTION

ALL MY LIFE I'VE WANTED TO SEE A GHOST. OVER THE YEARS I'VE BEEN IN ALL THE right sorts of places for a haunting – ruined monasteries, gloomy stately homes, ancient castles mired in history, lonely glens where the only sounds are birdsong and distant streams falling over stones, battlefields, abandoned cemeteries . . .

Sometimes I've gone ready to see ghosts and at others I've been distracted enough by work, or whatever, that I might have been taken unawares. But while I've had a couple of moments in a couple of places, when I've felt my hackles rise (of which, more later), I have yet to see anything I could describe as a ghost.

That said, I have had all manner of encounters with people who could say different: all sorts of people I've met, under all sorts of circumstances, have told me stories of things they've seen, and for which there seemed, to them, no rational explanation.

I do love a ghost story – and I'll make no bones about it, forgive the pun. I love the hearing and the telling. I love how ghost stories inhabit that part of us between the known and the unknown, between the certain and the uncertain.

Many years ago, while filming for a TV documentary about some or other bit of history, I got into conversation with a chap we'd invited along as an expert on part of the story we had to tell. He was a fellow Scot, middle-aged. I think he might have been an academic, but the truth is I can't remember. I wanted to put this story into the book, but no matter how hard I racked my brain, I couldn't come up with enough of the salient details – the place, the time, his name, the broader context. I think what he told me that day eclipsed the minutiae so that I have been left with the story and nothing more.

Filming days are long, with many periods of waiting for the crew to decide the conditions are right for more talking in front of the camera. During those times, and in those frankly strange circumstances (in the company of a relative stranger for hours on end), people unfamiliar with the filming process often offer unexpected titbits about their lives. It has to do with the need most people have to fill silences that might otherwise feel awkward. Eventually the small talk is done with and, to keep talking, someone sometimes must up the ante.

Over the years I've heard all sorts, about relationships, childhoods, loss . . . For reasons I can't completely explain, downtime spent with an interviewee can often turn into quite a confessional. I've wondered if, for some, it's an effect of my face being familiar to them. Maybe because they've seen me on telly, or on YouTube, they feel as if they know me well enough to confide in me.

We were in the Highlands of Scotland, this chap and I, on a stretch of wind-blasted hillside. Among other things, we were waiting for that wind to drop to see if the light might improve. He and I were some distance from the camera crew and the director, hands in pockets, preserving warmth and killing time. He nodded in the direction in which he was facing and said that, some years before, he'd been driving alone on a road a few miles distant. It was at night, in the middle of winter, in darkness and heavy rain. His windscreen wipers were making heavy work of it.

In the beam of his headlights he noticed, on the road ahead, someone walking. As he drew closer, he could see it was a young man in clothing less than suited to the conditions – a light jacket and jeans. He was soaked through. My man drew up alongside and leaned over to wind down the window on the passenger side – it was long enough ago that electric windows were not yet the commonplace they are now. The young man, shaggy-haired and dark-bearded, turned and leaned inside. A lift was offered and accepted – the young man would be happy to go wherever my man was headed, he said, to any civilization. He opened the door and climbed inside. He was drenched. Water was running off him and soaking the seat he sat upon. No matter, my man thought, he wouldn't have left a dog out on such a night. He put the car into gear and headed off.

There was some conversation at first, an acknowledgement of the miserable nature of the night, but not much. After a few minutes, silence fell, and the miles started passing. It was a peaceable atmosphere, though, no sense of threat or anything untoward. Given the prevailing conditions, my man had, anyway, to keep his eyes and his mind on the road for the most part, nearly hypnotized by the waves of rain, the ticking of the wipers, the beams of his headlights slicing through the dark. It was as much as he could do to follow the road, and the task demanded his concentration. After a few miles he felt perhaps the silence had gone on long enough and turned to ask the young man how he had come to be walking there at all – so far from any vehicle.

And no one was there. The young man had vanished as though he had never been. The passenger seat was quite dry.

I had and have no reason whatever to disbelieve a word of what my man told me that day. For his own part, surely he had no reason to make up such a story for someone he had never previously met. I remember he kept his hands in his pockets the whole time. His tone was calm and matter-of-fact, no need for added drama. I shook my head at the revelation of the empty seat. I remember making some remarks about being amazed, dumbstruck.

'I know,' he said. 'Me too.'

I live in Stirling, near the city centre. A few miles away there is a suburb called Cambuskenneth. 'Cambus' refers to a creek, or perhaps a haven. The Kenneth in question is King Kenneth I, usually described, by those who don't know any better, as the first king of Scotland, who defeated an army of Picts there or thereabouts. As it winds across the flat floodplain, the Carse, the River Forth seems at its laziest, its many drunken meanders making huge silvered loops in the grass as it all but loses it way. At Cambuskenneth the coiling almost makes an island of a great swatch of green, and there, in the twelfth century, another king, David I, gave land to a community of Arrouaisian priests. The ruins of their abbey are there yet, for the most part just a few courses of masonry, but the campanile – the bell tower – is much as they left it. Nowadays Cambuskenneth is a little community all of its own, the houses within easy walking distance of the city but far enough away to give a feeling of a place apart.

Over the years my daughter and then one of my sons took piano lessons in a house there. While they practised their scales and such, I would often wander around the ruined abbey. It's good, well worth a visit. I almost always had the place to myself. As well as the bell tower and the floor plan made of dark stones, there's a tomb containing the bones of a third king, James III, and those of his wife and queen, Margaret of Denmark. Margaret died first, in 1486. Then, in 1488, James was killed in battle at nearby Sauchieburn, in a scandalous betrayal likely involving his young son, who would succeed him as James IV and, legend has it, wear a hair shirt or perhaps a cilice ever after, as penance for his patricide. James and Margaret were placed together in the tomb. It was damaged during the dissolution of the abbey and restored in 1865 at the request of Queen Victoria.

When the light is low in the sky at the end of the day, Cambuskenneth Abbey feels like the perfect place for ghosts. But I've seen and felt nothing there at all. It's atmospheric for sure, the river close by, gravestones and the shade of trees. Everything about it feels like a stage set for haunting but, for me . . . nothing so far. That said, I've heard all manner of stories from residents of the nearby houses. Sometimes when I'm sitting in my car outside the house of piano lessons, people see me and come up to talk. Over the years I've heard multiple accounts of the sort of encounters I've longed for. Most recently – just a year or so ago – a couple stopped. He was in a mobility scooter and had to keep moving to let cars pass by before returning to his position beside my open window. His wife chatted too, agreeing with all he had to tell. He gave me a neat summary of Cambuskenneth hauntings – some I'd heard before and others I hadn't. There are several stories of encounters with ghostly monks and priests, figures darkly robed and hooded. Sometimes they are glimpsed in gardens, sometimes inside the houses. He told me about neighbours who'd looked up from their dinner table more than once to see a hooded face peering in at their window. There was a figure in a corner of a room, and another encounter outside, with a figure who disappeared into a hedge. One story I hadn't heard before, which was different from all the others, was about a sudden appearance, inside the man's friend's house, of an indistinct shape that whirled in space, like a dancing dervish, before disappearing as inexplicably as it had come.

Like the man on the hillside in the Highlands, he was completely calm and reasonable as he listed the stories. His wife, evidently well versed in them, turned away more than once to pass the time of day with passers-by. He had no reason whatever to make any of it up and I say, hand on heart, I had no reason to do anything other than take him at his word. He had not seen anything himself, but he vouched for the honesty of those from whom he'd heard the stories.

I think about those stories, and others besides, and wonder more than anything else about our apparent need for hauntings, and what that must say about us. Ghost stories have been with us since the beginning, so one might say their telling must express some part of the human condition. What follows is my account of more of the stories I've heard, or read, in places I've visited. In the end, all I can honestly say is that I just don't know what to think.

The book is also, in some small way, about my dad. When the idea for a story of ghosts came along, he was still alive. By the time I came to start writing it, he wasn't. His dying changed the book from the one I had planned to write into what it became.

Most of all, when I think about my dad, I think about his hands.

INTRODUCTION

For much of the time when he was alive, I didn't think about them at all. I don't think I even noticed them very often. If pushed, I might have noted that his hands were bigger than mine, his fingers thicker. During the weeks when he was in his bed and dying, I held them more than I had for years, perhaps ever. I remember the feel of them now as I type these lines. His hands seemed huge at the end, the skin cool and dry. Most of all they felt heavy, ponderously so. I have wondered since if time had him firmly in its grasp, reeling him in, pulling him like a stronger gravity.

This book is therefore inadvertently haunted, in its own right, by the memory of my dad, as am I.

1

Sandwood Bay, near Cape Wrath

A MILE SOUTH-WEST OF SANDWOOD BAY, IN CAPE WRATH, STANDS A WATCHMAN.
He is Am Buachaille – the Shepherd – a sea stack of Torridonian sandstone revered by climbers. Before any ascent, they must wait for low tide and swim thirty yards to his flat, pale plinth. From the shadow of the obelisk, like Lego bricks assembled by a giant's child, there is a glimpse, beyond the waves, of a fingernail-clipping of pink-gold beach. Whichever way you look at Sandwood Bay (from *Sandvatn*, which is Old Norse for 'sand water'), it is remote. The closest a car will bring you is the nearby hamlet of Blairmore, by Kinlochbervie. From there, it is a little over four miles' walk along a path across brackish moor to the rim of the North Atlantic. Even those who live nearby and love the place concede that the way to the beach is often bleak, barren. The path floods in patches. There are no trees and so, often as not, no birdsong, just quiet pressing down from above. Behind high dunes lies Loch Sandwood, thick with trout, and on its northern shoreline the roofless, ruined shell of Sandwood Cottage.

Around the wider landscape, subsumed by lonely years, there are homesteads that were emptied in the nineteenth century by the Clearances that made way for sheep. Stubborn walls are out-size planters now, for berry-laden rowans, scrubby shrubs and other overgrowth. As is often the way of landscapes that were home to people, then not, there is an almost palpable sadness made of their leaving. A landscape cannot mourn the missing but for any making the trip now across the machair – fertile, wide and deep, once cradling lives – the feeling of desolation leaves a sense of sentences unfinished.

There were Picts at Sandwood and thereabouts – the painted people the Romans met and clashed with further south. Their ties trailed back to the hunters who followed the deer that came and went as the last Ice Age waned ten thousand years ago. Then Scots from Ireland, more characters in a long roll call of tribes laying claim, then leaving. The Vikings, who left the name a thousand years ago, are said to have pulled their ships ashore and portaged them over the dunes into the loch beyond. Maps made in the 1600s warned travellers of roaming wolves. Residents now are few and far between. Sandwood and its environs are in the care of the trust named for John Muir, a nineteenth-century lover of wild and empty places. It is care that is needed, attention paid, if a place so vulnerable to violation – modernity and its ills – is to prevail.

Cape Wrath (more Norse, 'wrath' being a corruption of *havrf*, which means 'turning point'). Other, older, words are reminders of other, older, people. On the edges of the world many hands have groped and scrabbled for a hold. On their journeys between their homelands and Scotland's westernmost coastlines, the cape was a corner to be turned. Many longships foundered among the rest, serpent necks snapped, only some of countless vessels that came to grief out of sight and out of mind. Fishing boats, cargo ships, refugees from Spanish Philip's Armada were sent all to Hell by God in 1588: 'He blew with His winds and they were scattered . . .'

Some of all the lost, from those days to these, lie swallowed by Sandwood Bay or thereabouts. A writer, Seton Gordon, came there when timber wrecks still lay upon it. In his book *Highways & Byways in the West Highlands*, he wrote: 'All of them are old tragedies . . . almost buried in the sand far above the reach of the highest tide.'

In such a graveyard it may be no surprise to find stories of lost men. They're not all about men, though, the stories. In the first year of the twentieth century a shepherd, Alexander Gunn, came to Sandwood in search of a lost sheep. His collie was by his side. All at once the dog dropped at his feet, cowered and howled. Gunn followed its line of sight and saw, on a flat rock just proud of the sea, an unexpected shape. Not a seal, a woman more like, at least in part, and lying stretched without a care. Blonde hair made dark with wetness, a great beauty, sallow skin and green eyes. Whatever the truth, the form above the reach of the waves was, he guessed, no less than seven feet tall. Until his death in 1944 he would insist that he had seen a mermaid.

Down through the years and centuries there have been all manner of sightings of all manner of things at Sandwood Bay. Most concern sailors – or their wraiths. There is a recurring theme of scale, grand scale, like that of Gunn's mermaid, of seamen six and a half or seven feet tall. Great swells of men, heavy-set in reefer jackets and the like, hats and heavy boots. Some locals remember knowing or hearing about an Australian man who came to Cape Wrath and fell in love with a local girl. Sandwood Bay was where they walked together until life took him away from her and from there. It was only in spirit that he returned to wonder what might have been.

His is also one of the spectres that has been seen or felt in the ruined cottage by the loch, but not the only one. In his *Gazetteer of Scottish Ghosts*, Peter Underwood (1923–2014) described a trying night two hikers spent there, disturbed by what they said was the sound of a skittish horse, whinnying and stamping on the boards above, open to the sky.

Underwood was a ghost hunter of note and recorded, too, the experiences of an Edinburgh woman. Although she never visited Sandwood, she had a splinter of wood scavenged from the cottage's internal stairs that she kept in a little box. She had homes in Edinburgh and in London and inexplicable experiences in both – heavy footsteps at night when she was alone, crockery thrown by invisible hands. Once, in her home in the Scottish capital, she smelt pipe smoke and whisky. On turning to spy the source, she came face to face with a bearded sailor who looked her up and down, then left via a window.

Cape Wrath lighthouse, in the parish of Durness, was built by Robert Stevenson for fourteen thousand pounds and lit for the first time on Christmas Day in 1828. Its lofty perch, on cliffs more than four hundred feet above the sea, is sometimes swallowed by cloud. Its light – first paraffin, then mercury vapour and finally electric – made the navigating of Cape Wrath safer for mariners. Before its time ships foundered often, and many corpses were washed up cold and seaweed-wrapped on Sandwood Bay. The

And the light shineth in darkness . . . the turning at Cape Wrath

bones of men are buried there unmarked and unremembered so it is a place as sad as it is lovely.

From time to time there has been malevolence as well, or just an absence of welcome. Underwood records the experience of a father and son collecting driftwood for their home fire. Their pony was laden with the bounty of some hours' effort and they were about to leave when the beast, normally a placid creature, whickered and shied with alarm. Beside them, out of nowhere, stood a big man, bearded, eyes blazing. Pointing at the bundles of sticks he thundered: 'Take your hands off what does not belong to you – and leave my property!'

Unnerved – not least by the inexplicable manner of the other's arrival – father and son did as they were told, unloaded the wood and hurried away.

(Angry defence of property or place is a recurrent theme of encounters with ghosts. On the Orkney island of Rousay, in 1911, a farmer told of how he had been digging into a burial mound, in search of building stone, when there appeared at his side an old man, bearded, and in strange, tattered clothes: 'Thou are working thy own ruin, believe me, fellow, for if thou does any more work, thou will regret it when it is too late. Take my word, fellow, stop working in my house, for if thou doesn't, mark my word, fellow, if thou takes another shuleful, mark my word, thou will have six of thy cattle dying in thy corn-yard at one time. And if thou goes on doing any more work, fellow – mark my word, fellow, thou will then have six funerals from the house, fellow; does thou mark my word; good day, fellow.'

Rattled but unheeding, the farmer kept about his work, for while the ghost had gone his need of stone remained. Accordingly, in the days that followed, six of his cattle died and he had to arrange funerals for six people close to him.)

Over and over come the reports of unease and disturbance at Sandwood – of fishermen, shepherds and crofters taking a night's shelter in the cottage and waking with a start. Bumps in the night, faces at windows, heavy footsteps, anxious investigations by candlelight, upstairs and down, nothing to be found and no sleep to be had. Whoever he is, the big man leaves no trace. Visitors tell of seeing him at a distance, alone on the dunes, or by one or other gable end of Sandwood Cottage.

Andrew Green (1927–2004), another ghost hunter and a contemporary of Underwood, recorded a report made in 1967 by two

English women on holiday. They spotted a figure of towering shape by the ruin and walked towards him. He was dressed like a sailor, they said, heavy clothes, hat and boots. When they reached the cottage he was gone, no footprints in the soft ground all about. Again and again there are stories of encounters with a heavy man who walks on sand yet leaves not a mark.

The saddest tale by far at Sandwood concerns another father and his son. Caught out by a storm, three fishing boats from the Western Isles huddled together in some nook out of harm's way. For reasons unknown, in a lull, one boat put up its sails and made a run for home. Fisher custom dictated that when one set sail, the rest should follow, and so it was. The storm, even angrier than before, found them again and they were tossed like flotsam, sails shredded. One of the three fell away, out of sight. The skippers of the other two vessels saw a light (from a house kept up late by a woman giving birth) and found their way to a safe haven. The lone vessel – crewed by a father and his sixteen-year-old boy – was not so lucky. The boat was overwhelmed, and when the bodies were found later, in Sandwood Bay, they were lashed together by a length of rope. The father had seen to it, in his desperation at the end, that he would not lose his flesh and blood even in death.

As Dylan Thomas had it, in *Under Milk Wood*: 'Dusk is drowned forever until tomorrow. It is all at once night now . . . the lights of the lamps in the windows call back the day and the dead that have run away to sea.'

So, Sandwood Bay is haunted by stories. Stories of Spanish gold spilled on to the sands from shivered timbers, stories of love lost and remembered, stories of men lost and displeased by those they find in the bay, strangers to them and not those sought. Stories of fathers and their sons and hopes of ties that might hold fast on one side of death or the other.

Father and son. Inability or unwillingness to let go. The action of a father binding himself to his boy, with rope sea-soaked, in the teeth of a tempest, might be read in more than one way: a practical measure in hope of preserving life, but also a determination to avoid parting even in death. A last desperate hope they might be found together, even if the worst happened, and returned to loved ones if only in death. Which of them drowned first, elder or younger? Or were they overcome together?

Fisherfolk are made especially aware of life's fragility, that no line tied is strong enough. In years gone by they did not learn to swim – the inability fostering respect for the sea, commitment to the boat. In any event, fishermen, and their families at home, knew what to dread or to expect when a vessel left harbour. Long-lingering, too, the idea that those making their living from the land were ever hard-hearted towards those making theirs from the sea.

All around the coast, from Cape Wrath to Cornwall and back again, there are legends of wreckers who set false lights to lure boats on to rocks so their cargoes might be harvested from the shallows. Darker tales of corpses stripped of valuables, of struggling survivors put to death and plundered rather than anyone being allowed to live to report the crime. True or false, malicious myth or honest testimony, the notion of murderous landlubbers, men and women, making prey of their fellows, led novelist Charles Kingsley to write in *Prose Idylls, New and Old*, in 1849: 'Significant, how an agricultural people is generally as cruel to wrecked seamen as a fishing one is merciful.'

Thoughts of sailors and fishermen come to grief far from home might carry a cargo of unclaimed guilt. Bodies found and buried behind dunes without their names, thoughts of loved ones elsewhere left wondering what had befallen their menfolk. A bay like Sandwood, with its burden of the lost, might harbour thoughts of broken, bloated bodies stripped naked by rocks and the sea, hollowed eyes, sand-pocked flesh picked at by crabs and birds and crawling things, and hurriedly put out of sight by those who neither knew them nor truly cared. There are no tales of encounters with that sad dyad – the father and his son – only and always a man alone. Does he stand for all those lost there? Recurrent is the notion he might be looking for something (someone) lost, but whomever he meets is never the one he seeks.

2

AUGHRIM BATTLEFIELD, COUNTY GALWAY

'It is at Aughrim of the slaughter where they are to be found,
their damp bones lying un-coffined.'

TUIREAMH SHOMHAIRLE MHIC DHOMHNAILL,
SÉAMAS DALL MAC CUARTA

THIRTY-FIVE YEARS AFTER AUGHRIM (12 JULY 1691), THE BLOODIEST BATTLE IN
Ireland, barrels of bones freshly dug from the field sat on
Custom House Quay in Dublin for export to England and
grinding into fertilizer. A doctor-botanist, Caleb Threlkeld, saw
green growth on fleshless crowns. He called it, '*muscus innatus
cranio humano*'. Moss growing on a dead man's skull. The stuff
of life is ever on the move, out of sight and invisible, essence,
motes of existence transported from somewhere to elsewhere.
Seeds ride the wind. Beasts herded into towns and cities bring
more on their hoofs and in their guts. Where they leave their
muck, stowaways within take root on soil made fertile. Birds

follow, and bugs, so that a mixing takes place of the urban and the wild.

Threlkeld followed, too, making notes of all that had found a nook and thrived, and where. On Dublin's streets, in cracks in the pavement, on waste ground, he saw immigrants from farmers' fields – scarlet pimpernel and common eyebright, cow parsley and white clover, ribwort plantain, lady's smock, meadowsweet and wild angelica, wood avens and hogweed. He was especially enthralled, though, by those passengers hitching a ride from battlefield graves to riverside dock. Then, 1726, there was no shortage of dead men's bones in Irish soil, nor would there ever be.

If the powdered bone made fields fruitful, dead boys and men turned into food, the moss that clung made its own magic: 'The *unguentum armorium*, or weapon salve, is compounded of this.' The unguent, the ointment, was prized – applied to wound and weapon in hope of cures for battle's hurts. The more violent the death that left behind the bones, the more potent the salve that resulted from the green that grew up.

(What must become of the souls of those dishonoured dead when the trumpet sounds on the Day of Judgment? How will they find their scattered bodies, ready to stand and give account of themselves, when all that was left of them was transformed into food and so made parts of others?)

The fields of Aughrim saw the end of the Williamite War in Ireland, fought between armies led by Catholic James II – chased off his three thrones by the revolution of 1688, which his foes called glorious – and by Protestant William, Stadtholder of the Netherlands, husband to James's daughter Mary.

If it was the Williamite War by the end, the trouble in Ireland started long before. There are grounds for saying it all began in the twelfth century when Derbforgaill, wife of Tiernan O'Rourke, King of Breifne, ran off with or was abducted by Dermod Mac-Murrough, King of Leinster. MacMurrough was deposed for his behaviour by Ruaidrí Ua Conchobair, the High King of Ireland, and went crying all the way to English Henry II, who turned a deaf ear. The slighted king struck a deal with Henry's rival, Richard de Clare, 2nd Earl of Pembroke, known as Strongbow, who invaded Ireland with a force of arms. Determined that Pembroke would not supersede him across the Irish Sea, Henry sent a larger force and so began the Norman lordship of Ireland, earning Mac-Murrough a place in infamy as the Irishman responsible for England's rapine. For succeeding centuries those English and their descendants helped themselves to land, were brutal in their suppression of uprising or complaint, made Gaelic a rebel tongue and set in train centuries of tyranny.

Early in the seventeenth century, the will of the English prevailed once more over that of Gaelic lords with other ideas. The island was run by Englishmen out of Dublin. Colonies of Protestants – English first, then Scottish – had been planted in Leinster, Munster and Ulster on turf confiscated from Catholic Irish. Salt was rubbed into wounds with the imposition of the Anglican faith, assuming for its own all Church property. Catholics were barred from government, and failure to attend Anglican services was punishable by fines. While Protestants in England were allying with Parliament in defiance of King Charles I, in 1641, Catholics in Ireland, though angry with their lot, pledged allegiance to him. Charles was defeated and later beheaded, of course, and Cromwell had his Commonwealth. In Ireland, warts and all, he bloodily re-established Parliament's control and Catholic lords were dispossessed.

When Cromwell died, his Commonwealth with him, Charles II was returned to the throne. In Ireland Catholic hopes were high that their loyalty to his father would be repaid with the return of their lands. But the new king was dependent upon the same New Model Army that had defeated his father, and secured its support with gifts of confiscated Irish lands. Irish Catholics noted, bitterly, that they were punished for their loyalty to the Crown while Protestants were rewarded for rebellion against it.

After the second Charles, the second James, a Catholic, in 1685 succeeded to the throne. He left Irish land in English and Protestant hands but let Catholics back into politics. He made Richard Talbot, Earl of Tyrconnell, Lord Deputy of Ireland. Hardly an Irishman, he was a son of an English family from the Pale – the swatch around Dublin most tightly controlled by the Crown. Under Tyrconnell, Catholics were allowed back into positions of power in government and joined the army's officer class. In 1687 James made plain there would be 'liberty of conscience' for all religions in his three kingdoms, but that brought disquiet among the Protestants: they feared the worst of watching Catholics assert themselves once more among non-conformists and Anglicans.

In England, Protestants in Parliament were appalled by the birth of James's heir and the prospect of a Catholic Stuart dynasty. Behind his back they coaxed William of Orange to sally forth from the Netherlands and seize the English throne. Soon after his arrival, with an army of his own, English soldiers rallied to him. Pausing only to cast the Great Seal, stamp of his authority, into the Thames at Whitehall, James fled into exile in France.

Assuming he had abdicated, the parliaments of England and Scotland accepted William as their king and his wife, Mary, as queen.

The long island of England, Wales and Scotland was one thing, Ireland quite another. Catholics were in the majority and controlled the mass of fighting men. Catholic and Protestant had something in common: both understood their destinies depended on having one of their own upon the throne. Those loyal to James – Jacobites, from 'Jacobus', the Latin for James – moved to secure strongholds and strongpoints.

Only Protestant Derry held firm for William, and for a while it seemed the Jacobites would prevail. James arrived in March 1689. A Jacobite Parliament met from May to July. Catholic lands were returned and discrimination against the faith was ended. There was half-hearted skirmishing between Jacobite and Williamite but nothing that counted for much in terms of settling the matter. The Jacobite Parliament struggled always for funds and eventually took to paying the soldiers with coins made of brass. Nationalism was in the air, a whiff of independence. Understanding how much Ireland mattered now, William came to Carrickfergus in June 1690.

As well as Tyrconnell for the king, there was Major General Patrick Sarsfield, governor of the western province, who became Earl of Lucan. When William marched to besiege Limerick, it was Sarsfield who foiled his plans by leading cavalry to cut the noose.

William marched his army south, towards Dublin, until James's Jacobites crossed his line at the River Boyne, west of the town of Drogheda. There was fighting on 1 July and a Williamite win that was less than conclusive. Nonetheless James lost his nerve and, thinking he had lost Ireland and the war, took ship for France.

By January 1691 Tyrconnell had funds from France and the makings for more of a fight. A French fleet sailed up the Shannon in May. Aboard was Lieutenant General Charles Chalmot de Saint-Ruhe, come to take command of the Irish Army. Godert de Ginkel came to Athlone on 19 June with eighteen thousand men tied to the Williamite cause. The Shannon lay between the opposing forces, and the Irish, on the Connaught side, broke two arches of the bridge to keep their foe at bay. Ginkel had his men try to span the gap with wooden planks, but squads of Irish gave their lives to throw them into the river. Ginkel had word of shallows close by so a force crossed there. Soon after, they seized the bridge, from the Irish side, and the town of Athlone as well. Burning with shame at the loss, Saint-Ruhe pulled away five miles to the village of Aughrim, determined to make his stand and save his name.

Military man that he was, and capable, he had his men take up a position on a north–south ridge, at least a mile long, on a hill there called Kilcommedan. A marsh on the ground three hundred feet below made the ridge doubly secure and the Irish felt safe, with a pass through the morass, called Urachree, on their right flank, and the ruin of Aughrim Castle on their left. Ginkel approached, with twenty-five thousand men and cavalry, outnumbering Saint-Ruhe by perhaps ten thousand. As well as Englishmen, with Ginkel there were men rounded up from seven nations whose leaders opposed King Louis XIV. As well as Irishmen, there were Scots and Danes, Dutchmen, Huguenots and more, and far more heavy guns than Saint-Ruhe possessed.

The fighting took place on 12 July, feints and testing of defences either side, from midday until late afternoon. Saint-Ruhe's angst after Athlone was made all the worse by the presence of Sarsfield,

Eachdhroim an áir . . . *Aughrim of the slaughter*

who had seen him fail, so he sent his second to the rear, out of sight, and never made him privy to his battle plans. It was a fatal mistake, given all that would unfold.

Ginkel had men wade, chest-deep, into the morass and on to the slope. Yard by yard they fought their way uphill against heavy musketry, and countercharges from Irish soldiers, who could take cover behind parallel rows of whitethorn hedges on the slope. Drawn too far forward of their own lines, Ginkel's regiments lost their nerve, broke and ran, plunging all the way back to the morass and beyond. The Irish cavalry pursued them, felling men like lumber. Yet more English came on, attacking the Irish close to Aughrim village. They, too, were turned back and, seeing the English panic, Saint-Ruhe cried, 'The day is ours, *mes enfants!*'

His excitement was premature. Ginkel regrouped his men. A force was sent against the Irish closest to Aughrim Castle and there the defenders found, to their consternation, that their musket balls had been cast for English weapons of larger calibre. Since they would not fit into their slender barrels, the Irish tore the spherical metal buttons from their tunics and loaded those instead. Their effect was negligible, and now the fortunes of war began to turn in favour of Ginkel. Still undaunted, his blood still up, Saint-Ruhe on horseback put himself to the fore of a cavalry brigade. Witnesses said he was adorned in all his finery, his medals on his breast, rightly attired for a victory that would tear Ireland from William's hands. He drew his sword and held it high, shouting to those around him that it was time to chase the enemy all the way to Dublin. So saying he kicked his mount into the charge. Those cavalrymen around him answered in kind, but the sound of his voice still hung in the air when a cannonball from an English gun parted his head from his body. His headless corpse, in glittering uniform, stayed in place upon its mount, first monument to a lost cause.

The shock was limited at first to those who saw it, word not travelling as fast as the caput rendered kaput. Still the Irish pressed their advantage and, had Sarsfield been better briefed of tactics, might have carried the day without Saint-Ruhe. Their enemy, though, sensing something turning in their favour, pushed forward on their own account. Those Irish who had seen or heard what had happened to Saint-Ruhe had been first to panic and depart. All through the Irish lines the malaise spread until there was a rout. Cavalry did best in their departure, of course. *Sauve qui peut*. Infantry did worst. Having thrown away their muskets to run unburdened, they were overtaken and cut down in vast numbers.

Two regiments stayed put and together, as darkness fell about them, in a cleft called Gleann-na-Fola. Next morning they were found in their defiance and butchered to a man. All in all it was a slaughter of the routed. If Ginkel lost three thousand in the fight, the Jacobites lost four or more, cut down as they fled. Numbers counted in the aftermath of such days are always lies but heaps of slain lay all about.

Ginkel made some effort to bury those who had fallen for him that day, but dead Jacobites were left where they lay, carrion. Since the landscape had been emptied of people there were no Irish who cared enough to do otherwise. Legend has it that a greyhound belonging to an Irish officer stood guard over his master's corpse. Stripped by the victors, like the rest, still his guardian remained. Other corpses were torn and consumed by scavengers, dogs among them, but the greyhound would let no such thing happen to his own, as best he was able. By January of the following year the officer was bones and threads of flesh, hair for lining birds' nests, but still his hound stayed by him. Eventually an English officer wandered on to the field and, challenged by the hound, shot it dead.

If the fallen were left above ground, the thought of them burrowed deep into Irish thought. *Eachdhroim an áir*, they called it – Aughrim of the slaughter. These were the poor people, the hospitable people Threlkeld would write about, upon whose bones would grow, in time, green moss. As well as seven thousand men, some dreams were felled that day – of a Catholic king, or Irish independence, or the righting of old wrongs. Unquiet ghosts, every one. Jacobites would rise elsewhere, and fall again.

Eyewitnesses after Aughrim said the dead were spread for four miles, or as far as the eye could see, and that they had the look of a vast and forlorn flock of sheep roughly shorn. The stump of Aughrim Castle is still there, wrapped in ivy and casting its shadow on nondescript pasture, seeming hardly cup enough to hold the tale without spilling.

In *Teague Land: or a Merry Ramble to the Wild Irish*, the Englishman John Dunton wrote in 1698: 'After the battle the English did not tarry to bury any of the dead but their own, and left those of the enemy exposed to the fowls of the air, for the country was then so uninhabited that there were not hands to inter them . . .'

Shamed by his hand in the death of Christ, Judas Iscariot had returned his thirty pieces of silver to the Temple authorities who had bought his betrayal. Knowing it for blood money, they would not keep it either so used it for the purchase of Haqel D'ma, Aceldama, some acres of waste ground pitted by the extraction of clay for bricks and pots, and suitable only for the burial of strangers and the unclean. Dunton labelled Aughrim the same, a Potter's Field.

It is awful – from Old English *egeful*, 'worthy of respect or fear, striking with awe, causing dread' – to go to Aughrim knowing what happened. A battlefield is ever a strange destination, after earth has stopped the ears of the dead. Regardless of any opinion about the politics and passions, it was a killing field for the butchering of boys and men. Go there in sunshine or in shadow and it is hard to chase off sombre thoughts of suffering, of yet more poor folk pulled into the grinder of rich men's machinations. Visitors report seeing soldiers standing still, always still, and always looking off into the distance. Mistake them for fellow tourists, walk towards them . . . and they drift away into the invisible. There is a nondescript depression close by the castle called 'Bloody Hollow' – as there is on every other battlefield, recurring tribute to memory of the fiercest fighting – where green was turned red by all the bloodletting. There and elsewhere stories are told of a sense of dread or inexpressible sadness. Some speak of the clutching touch of unseen hands. Talk of agonized cries hanging in the air is common.

Most poignant are reports of the officer's hound, still waiting, still holding ground. By definition, loyalty lasts. Loyalty is the long game. Do people coming to Aughrim bring their own thoughts of fidelity, of keeping or breaking faith with a dream or a cause? Do they find fidelity's best manifestation in thoughts of a hound that will not break faith, even unto death?

As at Sandwood Bay and elsewhere there is cause for guilt at Aughrim, in the hearts of any who visit. Deuteronomy 24:16 has it that no child shall be put to death for the sins of the fathers, which seems about right. But knowing of the casual ending of the ancestors leaves a sour taste. Walk the fields of Aughrim, flinch and raise a shoulder to fend off thoughts of others taken young through no fault of their own. As Humbert Wolfe wrote in 'Requiem: the Soldier'

Down some cold field in a world uncharted
the young seek each other with questioning eyes.
They question each other, the young, the golden-hearted,
of the world that they were robbed of in their quiet paradise.

According to *signa naturae*, the Doctrine of Signatures, God sets disease and cure side by side. Stinging nettles here and soothing dock leaves there, as our mothers taught us. In the case of *muscus innatus cranio humano*, the ointment from the moss might cure ills of the head, of the bones. Paracelsus – 1493–1541 – noted that '*sola dosis facit venemum*', the dose is the poison. Anything and everything is toxic, he reasoned. Too much air, too much water, whatever, it was just a question of how much of that thing was taken. In Ireland there has been a toxic dose of strife, the ague in the blood of every man and woman, and while the infection did not start with the Williamite War that ended at Aughrim, it was a symptom of an especially bad outbreak of a sickness centuries old.

Every year, on 12 July, Orangemen march in Northern Ireland in memory and celebration of Protestant William's victory over Catholic James at the Boyne. The Battle of the Boyne was on 1 July, however, and originally it was Aughrim, on 12 July 1691, that Loyalists remembered. As the loved and loathed 'The Sash' – a traditional folk song has it,

> And on the Twelfth I love to wear
> The sash my father wore
> It's old but it is beautiful
> Its colours they are fine
> It was worn at Derry, Aughrim,
> Enniskillen and the Boyne . . .

A younger wound from an older anger, Aughrim ended nothing. Like ghosts and talk of ghosts, hatreds linger, passed down through generations, believed and not. The battlefield, by Aughrim village, between the towns of Loughrea and Ballinasloe, on what was once the main road between Galway and Dublin, was farmland then and still is. There are two ringforts on the field (like ripples frozen after two pebbles are dropped into a pond), defended farmsteads raised against others, each other, older trouble. One is Aughrim Fort, known also as Saint-Ruhe's since it was behind its rampart that the Frenchman had his guns. The other is Lisbeg, and both command fine views towards the Melehan River.

Grudges, blood feud, unfocused hatreds made airborne, hate made genetic, like a Viking Claw or a kiss curl. No one travels empty-handed and visitors to Aughrim bring their own thoughts at least. The cries of dying men, desperate hands clutching at a trouser cuff. A spectral hound glimpsed out of the corner of an eye. Is it any wonder?

3

LADYBOWER RESERVOIR AND THE DAMBUSTERS, DERBYSHIRE

'Day is dying in the west;
Heav'n is touching earth with rest;
Wait and worship while the night
Sets her evening lamps alight
Through all the sky.'

MARY A. LATHBURY, 'DAY IS DYING IN THE WEST'

MY DAD DID HIS NATIONAL SERVICE IN THE RAF. IN A FAMILY ALBUM, somewhere in my childhood home, there is a photograph I see only in my mind's eye. It was taken near Changi Air Base, in Singapore, in the 1950s, where he spent most of his three years. I could go looking for it and check the details, but I'm happy with the image I carry in my head, half remembered. So handsome to me, my dad. Brylcreemed hair. Just a boy, eighteen or nineteen, a little older or younger than my own children. So young his features need some growing into. I am sure he has shorts on in the shot, making him seem even younger. He has a cigarette in one hand. For all my growing up and into my thirties, Dad was an enthusiastic smoker. Sometimes a pipe, sometimes those Café Crème cigars that came in a little flat tin. Mostly it was cigarettes. I was often sent to buy them, back in the days when shops did not seem to mind selling such to children: 'Twenty Embassy Red, please, for my dad . . .'

More than once he said he regretted not staying in the RAF once his National Service was over, that it might have given him a career. From time to time, once in a blue moon, really, he would have an evening out at the local branch of the Royal Air Force Association wherever we were living – in Ayr, in Dumfries. I have his house keys now, with a brass button from his uniform, which he had repurposed as a fob. King's crown and eagle. I would put it on the ring alongside my own keys and carry it with me, but I am too afraid of losing it.

Those connections, slight as can be, made and make me pay more attention to the RAF than the other services, especially to movies featuring wartime derring-do of magnificent men in flying machines. *The Dam Busters*, scripted by R. C. Sherriff, was a favourite, and is still enough to make me sit down and watch whenever the past recedes like a tide, returning it to view. 'The Dam Busters March' . . . Richard Todd as Wing Commander Guy Gibson . . . Michael Redgrave as bouncing-bomb inventor Barnes Wallis . . .

'Is it true? All those fellows lost . . . fifty-six men . . . If I'd known it was going to be like this, I'd never have started . . .'

'You mustn't think that way . . .'

My mum knows I am writing about ghosts, and also about Dad. Not long after I told her about the book I planned to write, she gave me a page from the *Sunday Express*, dated 5 June 2022. Mum always reads the *Express*; Dad always read the *Daily Mail*. 'Ghostly Dambusters get into the fighting spirit . . . bombers are said to have returned to haunt the valley where their pilots trained – sparking claims of a "timeslip" phenomenon.'

The valley in question is Upper Derwent, at the heart of the Peak District, in Derbyshire. Between 1935 and 1943 it was dammed and flooded to create three reservoirs, providing water for the cities of Derby, Leicester, Nottingham and Sheffield. The 'ghostly Dambusters' were reported flying low over one of them, Ladybower, largest of the trio, which also includes Derwent and Howden. According to the newspaper article, a woman sitting in a car with her husband saw a bomber, dark and silent, flying 'quiet as a mouse' over the surface of the water. She watched until it seemed to disappear into the landscape before turning to her husband. 'Did you see that?'

Hers is not the only testimony of that sort. Another couple were by Ladybower on a moonlit night in recent times and reported the shadow of a Lancaster, sailing low over the silvered depths. In recent years there have been more glimpses of the same.

Long before the Lancasters, Ladybower had a reputation as a haunted, haunting place. Beneath the billions of litres of water two villages lie drowned. Compulsory purchase orders saw to the clearing of hundreds of residents of Ashopton and Derwent in the years before the Second World War. Ashopton, the larger, had grown, over uncounted centuries, around a crossroads on the main route between Sheffield and the market town of Glossop. As

Casting a shadow of past times – the seventieth anniversary of the Dambusters' Raid

befitted a crossroads, where travellers might pause, there was a coaching inn, the Ashopton, as well as a post office, general store, a garage and a Wesleyan Methodist chapel. There were scores of houses and farms. Every July, since a time before the reach of memory, there was a wool fair.

Derwent sat by the river of the same name. There was a manor house, Derwent Hall, built in 1672, for one Henry Balguy of The Hagg, of local stone, mullioned and transomed windows in the Jacobean style, fine gardens, and an ornamental fishpond. It was owned later by the dukes of Norfolk. In the years before the waters rose it was a youth hostel, then briefly a school. Elsewhere, in the village, winding streets were lined with cottages, a school, the Church of St John and St James, with space for 140 parishioners, raised first as a private chapel for the Balguy family. Surviving records show that the last marriage took place on 22 December 1938 when the minister, Mr Simpson, heard the vows of Miss Olive Ollerenshaw and Mr Frank Booth. The last service of any kind, a farewell to the place, was held on 17 March 1943, attended by the Bishop of Derby and officials from the Water Board. Ashopton's Methodist chapel fell silent for ever on 25 September 1939. The loved bodies of the dead from both churchyards had been exhumed and reburied among strangers in a graveyard in the village of Bamford. It is difficult to be sure, but the last hymn the Methodists sang in their chapel is said to have been 'Day Is Dying In the West', by Mary A. Lathbury.

When forever from our sight
pass the stars, the day, the night,
Lord of angels, on our eyes
let eternal morning rise,
And shadows end.

Construction of the necessary dams had been ongoing since 1899 when an Act of Parliament gave Derwent Valley Water Board the go-ahead for six reservoirs. Derwent and Howden dams were raised in the years before the First World War and from then the waters rose. Once Ladybower was completed, the rest of the valley took two more years to flood, the water creeping ever higher, insidious as doubt. The main body of St John and St James was demolished – like all the buildings in both villages – but its spire was left, at first, as a memorial. It stood defiant above the water's surface, an accusatory finger, for a year or so and then, on some or other safety grounds (there were tales of folk swimming out and climbing it) it was dynamited into oblivion. Tales of a ghostly ringing of the church bell on quiet nights are challenged: the bell in question was taken from the spire in 1943, and later hung at St Philip's Church, in Chaddesden, a new church built in 1955.

It was all a deliberate undoing, an unmaking of a place that had been home to generations uncounted. Not much of note may have happened there, but it might be assumed it was a few square miles that mattered to those who had known it as home, that had mattered to those laid to rest in spaces briefly borrowed from the clay in the churchyards.

A few Derwent houses and farms are there to this day – on higher ground beyond the water's reach. There is still a civil parish of Derwent, with a population counted at fewer than a hundred. One other structure survived the flood: a seventeenth-century packhorse bridge was removed, block by block, and rebuilt by Howden Reservoir at a site called Slippery Stones. It is now a Scheduled Monument.

That which is lost may return to view. Over the years, in times of drought, the level of Ladybower has fallen low enough to reveal ruins slick with slime. It happened in 2018 and again in 2022. Lost loved ones may take us by surprise, thoughts or seeming glimpses of them returning. We are left to wonder if they are really gone . . . or is our view obscured only by elements in between so that they are still there, just below the surface?

Everyone calls it the Dambusters' Raid, but it was codenamed Operation Chastise. An attack on the heavily industrialized Ruhr

Valley, in western Germany, was thought likely to cause significant disruption to the Nazi war effort. To that end, the Möhne,
Edersee and Sorpe dams were targeted. Since the enemy was
more than aware of their importance, not least because of their
capacity for generating hydroelectricity, all manner of defences
had been deployed, including nets in the water to stop torpedoes
reaching the dams. The surrounding terrain for miles around
bristled with anti-aircraft guns and searchlights.

In his own time, Admiral Horatio Nelson had reported cannon
balls bouncing across the sea towards his enemies' vessels. Suitably inspired, English engineer and inventor Barnes Neville
Wallis designed a twentieth-century upgrade. Slung half in and
half out of the belly of a modified Avro Lancaster, barrel-shaped
bombs were given backspin by an auxiliary motor before they
were loosed. The necessary low altitude for release was pinpointed
by two spotlights, carefully angled, fore and aft on the planes'
undersides. When those beams converged to form a perfect circle
on the water's surface the aircraft were at the ideal height of 60
feet. If all went well, the spinning bombs would bounce across the
surface of the reservoirs, leapfrogging the torpedo nets and anything else, then strike the dams. Their residual spin would drive
them beneath the surface to maximize the power of their explo-
sions underwater.

Nothing of the sort had ever been attempted so practice runs
were essential. By late 1943 Squadron X had been formed. It was
later numbered 617 and led by Wing Commander Guy Gibson,
just twenty-four years old but already a veteran of 170 bombing
raids elsewhere. His fellow fliers and crew were brought together
from all over the Allied territories – Britain, Australia, Canada,

New Zealand, the USA. They had just a month to prepare for the specific demands of Operation Chastise, and the reservoirs of Derwent Valley were among the sites selected for perfecting the necessary skills.

Overnight between 16 and 17 June 1943, 133 airmen in 19 Lancasters attempted to complete the task. The planes flew low through the enemy dark, the bombs bounced and the Möhne and Edersee dams were breached, causing hellish damage. As many as 1,600 people on the ground were swept to their deaths by the ensuing torrents, perhaps a thousand of them forced labourers. The Sorpe dam survived, but the success of Chastise was a mighty boost to the morale of Britons battered by the Blitz. There was significant disruption in the Ruhr area, although it was short-lived. When all was said and done, 53 members of 617 Squadron were dead, and 3 had been taken as prisoners of war. A total of 8 aircraft were lost. Gibson lived to tell the tale and was awarded the Victoria Cross. In all, 34 of the survivors were decorated for their bravery: 5 DSOs, 10 DFCs, 2 Conspicuous Gallantry medals and 11 DFMs. Historians still debate the effect of all that effort and sacrifice, with German writers often more inclined than their British counterparts to declare the Dambusters' Raid worthy of its luminous memory. Wallis later lamented that there was no attempt to capitalize on its success, no subsequent raids to disrupt German efforts to repair the dams and power stations. Arthur 'Bomber' Harris, in charge of Bomber Command, apparently regarded it as having been a waste of men and resources.

On 16 May 2008, Derwent Reservoir was the scene for commemoration of the sixty-fifth anniversary of the raid, with a flypast by Lancasters, Hurricanes and Spitfires. Squadron Leader John Leslie Munro, CNZM, DSO, QSO, DFC, of the Royal New Zealand Air Force, by then the last surviving Dambuster pilot, was there to watch. So was Barnes Wallis's daughter, Mary Stopes-Roe, and Richard Todd, representing the cast and crew of the movie.

Derwent Valley was a training ground for 617 Squadron – that much is certain. There are photographs aplenty of Lancasters flying low over the waters, seeming to skim the dams. In more recent times the planes have been back in their old haunt more than once. There has been, too, a continuity of memory – cords stretching from those men, young, who died in 1943 to those who remembered them and grew old. The valley doubled as the Ruhr for the purposes of the movie, prompting a return in the mid-1950s of more Avro Lancasters. They were modified like those used for the raid, with the removal of their mid-upper gun turrets, so that their silhouettes were the same as those of the originals. Glimpses linger, like smoke, or perfume, clinging to the fabric of time. More Lancasters came back to the valley in the 1990s as well.

It has been suggested those potent moments, broadcast on television news and captured in photographs, may have stuck deep in the consciousnesses of some and lingered, inspiring the imaginations of those witnesses subsequently reporting seeing 'ghostly Dambusters' where the real raiders once flew. Love is in the mix, and grief for those lost into darkness over Germany on that night of nights.

The mention of 'timeslips' in the newspaper article invokes the theory, usually the stuff of fiction, that individuals might travel by accident from their own time into some other. In the case of the ghostly Dambusters, the notion would have those aircraft briefly leaving 1943 and passing through subsequent years for mere moments before returning to their own time. In his novel *Contact,* cosmologist Carl Sagan had his protagonist, Eleanor Arroway, travel through a wormhole, a portal between her own twentieth century into a place and time in the distant future. Earlier in the story Arroway receives news of her father's death while reading *A Yankee in the Court of King Arthur.* Mark Twain's 1889 novel was one of the first to exploit a timeslip – to get his lead character from twentieth-century Connecticut to the sixth-century court of King Arthur.

Outside the realms of fiction, some are open to the concept that places may bear the 'emotional residue' left by the experiences of people who spent time there, as described by Savani, Kumar, Naidu and Dweck in their 2011 paper, 'Beliefs about Emotional Residue: The Idea That Emotions Leave a Trace in the Physical Environment'.

Celtic Christianity reveres the notion of 'thin places' in the landscape. In *Man Seeks God,* American author Eric Weiner described 'those rare locales where the distance between heaven and earth collapses', where the separation between dimensions seems gossamer thin. Scratch the surface of our publicly stated beliefs and we may find ourselves privately admitting to times in our own lives when logic and reason said one thing while our gut said another. English philosopher Roger Scruton wrote in his book *Confessions of a Heretic*: 'Anybody who goes through life with open mind and open heart will encounter these moments of revelation. Moments that are saturated with meaning, but whose meaning cannot be put into words.' Perhaps there are spaces between spaces, big enough even for Avro Lancasters to slip through.

Loss might leave a lasting residue. At the beginning of *Unreliable Memoirs*, the first volume of his autobiography, Clive James quoted lines from Homer's *Iliad*, concerning Andromache's grief at the loss of her husband, Hector, Troy's greatest warrior:

'Husband you are gone so young from life and leave me in your house a widow ... You leave woe unutterable and mourning to your parents, Hector, but in my heart above all others bitter anguish shall abide. Your arms were not stretched out to me as you lay dying. You spoke to me no living word that I might have pondered as my tears fell night and day.'

The book ends with James leaving the Australia of his childhood and early manhood, bound for England and Cambridge University. His father had died in the war, leaving his mother to raise their son alone. Ghosting through the book are glimpses of James's realization that he came late to noticing, far less understanding, his mother's loneliness, the toll taken on her by the loss of her husband. He admits boarding the ship bound for the British Isles, and his future, without so much as backward glance or a thought for his mother, left behind on the wharf, and in her past. Memoirs and memories: both are unreliable – but there is truth in the book. Earlier James writes about his first love, how years later he could still experience the intensity of his feelings then, even though he could no longer recall the face of the girl in question: 'We forget the shape of the light but remain dazzled for ever.'

Are we tricked by the intensity of our feelings, so dazzled by our memories that we see what was there once but no longer? The heart wants what the heart wants.

4

OLD VICARAGE, GRANTCHESTER, AND RUPERT BROOKE

THE FIRST WORLD WAR HARVESTED 20 MILLION MEN BUT SOMETHING ELSE DIED too, something older. Iron railings were set between past and present. That image is not mine, rather something I have read, but it is right enough. The past was still visible but forever out of reach. What died between 1914 and 1918 haunted the rest of the twentieth century, and still haunts the twenty-first. The pains around the world are from wounds inflicted then. My grand-fathers served and survived: Robert Miller Oliver and James Cameron Neill. Both bore wounds ever after – shell fragments for Robert and the tumbling track taken by a bullet for James. I have been haunted by the thought that either or both wounds might have seen to my non-existence. In *Requiem for a Nun*, William Faulkner said: 'The past is never dead. It's not even past.'

Anyone thinking the Great War is past is mistaken. I say we are the damaged descendants of damaged ancestors.

If a residue of emotions might hang in the air of a place, then English war poet Rupert Brooke imagined his own essence, his mortal remains, would be transformative wherever they lay. In the fifth of his War Sonnets, he wrote:

> If I should die, think only this of me:
> That there's some corner of a foreign field
> That is for ever England. There shall be
> In that rich earth a richer dust concealed;
> A dust whom England bore . . .

In fact, Brooke died at sea. He had been bound for the landings at Gallipoli, departing England with the British Expeditionary Force on 28 February 1915, but became ill with sunstroke and dysentery at Port Said, in Egypt. His constitution had always been weak, his immune system frail. A mosquito bite on his lip became infected, leading eventually to sepsis. He had been so debilitated he ought never to have left Egypt. He was aboard the French hospital ship *Duguay-Tourin* in a sheltered bay off the coast of the island of Skyros, in the Aegean, when his condition worsened and he succumbed to blood poisoning. He died in the late afternoon of 23 April – St George's Day and the birthday of Shakespeare. He was twenty-seven years old. The ship and the rest of the forces could not tarry, and Brooke was buried in the evening of the same day, in a grove of olive trees on Skyros, a mile from Homer's wine-dark sea.

He had been born into privilege – son of William, a housemaster at Rugby School, in Warwickshire, and Ruth, a school matron.

Taught first at home, by a tutor, he attended Hillbrow Preparatory School, then Rugby, where he excelled academically and at sport, including the ball game that takes the school's name. In 1904 his weak constitution resulted in a bout of illness, followed by a convalescent stay of some months with family friends in their villa near Genoa. He used the time to write, including poetry, for which he had already demonstrated an aptitude. He won a scholarship to Cambridge to read classics and English literature. After a year in college rooms, he left for digs at the Orchard tea rooms, owned by the Stevenson family, in leafy, dreamy Grantchester, three miles up the Cam. He attracted and was attracted to louche company, including Virginia Woolf, who styled themselves neo-pagans. They walked barefoot, favoured natural over urban landscapes, swam naked in rivers and pools, talked poetry and socialist politics.

After the death of his father in 1910, or thereabouts, Brooke became ill once more. When he had recovered, he moved to the nearby Old Vicarage. Described by the Irish poet William Butler Yeats as 'the most handsome young man in England', he was a magnetic and seductive presence wherever he went. Attracted to both women and men, he had intense relationships with one after another, consummated or not. While at Cambridge he was friends with E. M. Forster and Bertrand Russell, Lytton and James Strachey, Maynard Keynes and others. Soon enough he was counted among the Bloomsbury Group ('Lived in squares, loved in triangles', as Virginia Woolf had it), meeting Henry James and Winston Churchill along the way.

At times, the complexities and uncertainties of his love life seemed to get the better of him. He had a breakdown of sorts

A lost idyll . . . the world of before – The Old Vicarage, Grantchester

around the end of 1911. In the aftermath, he travelled. Homesick in Berlin, in 1912 he wrote 'The Old Vicarage, Grantchester'.

> And laughs the immortal river still
> Under the mill, under the mill . . .
> Stands the Church clock at ten to three?
> And is there honey still for tea?

It is a picture of innocence, even of naivety, certainly a paean to an England that only ever existed for a hallowed few and was soon, anyway, to disappear for everyone for ever. Onwards to the USA, then the South Pacific and Tahiti, where he fell in love with a local woman and fathered a daughter he would never know.

Hasten, hand in human hand,
Down the dark, the flowered way,
Along the whiteness of the sand . . .

Leaving that happiness behind (if true, uncomplicated happiness it was), he returned to England. Within months there was world war: he sought and was awarded a commission in the Royal Naval Volunteer Reserve. His group was posted first to Antwerp, in Belgium, and it was there that he composed the poems for which he is best known, the five War Sonnets called 'Nineteen-Fourteen'.

In those works there is yet more of what was to die in the war, alongside all those men and boys. In the first, 'Peace', Brooke imagines war as a relief for young men like him, grown weary of peace, of lives unchallenged and therefore without meaning.

Now, God be thanked Who has matched us with His hour,
And caught our youth, and wakened us from sleeping!
With hand made sure, clear eye, and sharpened power,
To turn, as swimmers into cleanness leaping,
Glad from a world grown old and cold and weary.

Naivety? Or blindness to reality? We cannot and should not judge. In any event the words are hard to read from this side of the railings, the sentiment unbearable given all that was to come . . . 'To turn, as swimmers into cleanness leaping': there was nothing clean about what lay ahead for millions of men, but 'Nineteen-Fourteen' made Brooke famous while he still lived. In the fourth of the five he described death as an infinite frost that leaves: '. . . a white/Unbroken glory, a gathered radiance'.

The notion of pristine perfection, as though those dead to come would be preserved ever after as they had been in life – instead of the crushed and torn reality of men's bodies rendered into tatters – was a seeming preoccupation of Brooke. His own death, indeed, had a Homeric quality, that most beautiful of young men, plucked in his prime as though by gods come to reclaim all that they had first bestowed upon him. That he was buried by moonlight – a volley of shots fired over his grave, a wooden cross to mark the spot – seemed to make real the poet's own prediction of a silvered eternity of peace.

Churchill sensed the power latent in his friend's fall at such a time, in such a place. He penned part of the obituary that was published in *The Times* on 26 April 1915, beneath opening lines that misreported his death as having been caused by sunstroke. Churchill wrote:

> . . . this life has closed at the moment when it seemed to have reached its springtime. A voice had become audible, a note had been struck, more true, more thrilling, more able to do justice to the nobility of our youth in arms engaged in this present war, than any other more able to express their thoughts of self-surrender . . .
>
> . . . Joyous, fearless, versatile, deeply instructed, with classic symmetry of mind and body, ruled by high undoubting purpose, he was all that one would wish England's noblest sons to be in the days when no sacrifice but the most precious is acceptable, and the most precious is that which is most freely proffered.

Gulled by these and other such sentiments, a generation marched into the meat grinder. Other men would die in other wars, but the innocence of those who trusted the idea that they had been woken by God to leap, as swimmers, into cleanness, would not survive the years between 1914 and 1918. The shade of their innocence then is some of all that haunts us now. Haunting us, too, is the ghost of the poet Brooke might have become had he lived longer, seen more of what lay ahead. Found in the notebook he was using before he died, 'Fragment' sketches images he gleaned aboard ship as the Hood Battalion made its way towards Gallipoli:

> Pride in their strength and in the weight and firmness
> And link'd beauty of bodies, and pity that
> This gay machine of splendour 'ld soon be broken
> Thought little of, pashed, scattered . . .
> Like coloured shadows, thinner than filmy glass . . .
> Perishing things and strange ghosts – soon to die
> To other ghosts – this one, or that, or I.

Brooke's own innocence was mortal, already moribund perhaps. By April 1915 he had watched the sacking of Antwerp. Bloodless though it was, it made Brooke witness to refugees streaming away from stolen lives, carrying whatever they could. It was the outer edge of horror. In any event, Brooke ought hardly to be blamed for not reflecting what he did not live long enough to see and therefore to know. Had he lived to encounter the trench warfare of the Western Front, he, too, might have acquired the bitter irony that suffuses the work of Wilfred Owen, and others, whose innocence was flayed from them, like living skin, to leave men made old before their time.

> Gas! Gas! Quick, boys! – An ecstasy of fumbling,
> Fitting the clumsy helmets just in time;
> But someone still was yelling out and stumbling
> And flound'ring like a man in fire or lime . . .

After Brooke's swimmers, into cleanness leaping, Owen saw enough – as veteran and author Paul Fussell observed in *The Great War and Modern Memory* – to transform them into 'the mud-flounderers of the Somme and Passchendaele sinking beneath the surface'. Whatever else Brooke was, whether golden boy or flawed human being, like those soon to be harvested in their millions, he immortalized the world he professed to know as effectively as anyone. As he died, so did that Elysium.

Back in Grantchester, the Old Vicarage has been owned, since 1979, by Jeffrey, Lord Archer, and his wife, Mary. What's theirs now is an elegant, stately building behind tall brick walls and imposing gates. Long before the Archers' tenure, there were reports of Brooke's shade walking in the garden, of books and other objects seen to move from shelves on the top floor of the building. Other visitors or tenants have reported hearing foot-steps around the outside of the house and inside walking towards the sitting room. From time to time a figure is seen in the house and in the garden, and close by the Orchard tea rooms, Brooke's first home in Grantchester. Most often those are attributed to his ghost. Nearby, on the Cam, is Byron's Pool, where Brooke's poetical predecessor used to swim. Others have claimed to encounter Byron's ghost there, glimpsed in the shaded waters, pale as a fish. Brooke was regarded by many, before and after his death, as Byron's heir – more gilded beauty of the same sort. While Brooke died en route to Gallipoli, Byron had swum the narrows of the Hellespont in 1810, more than a hundred years before. Both died in Greece, in April, in time of war – Byron at Messolonghi, in April 1824, while fighting the Ottoman Empire during the Greek War of Independence.

HAUNTINGS

The sadness made of the First World War is timeless and end-less. It is altogether too big, too grand, to accommodate in the mind. It meant and means too much. Its remembrance is made of people and places set aside. The awful cemeteries of Flanders where Brooke's vision was made real for millions, each man assigned his corner of a foreign field. The Cenotaph in Whitehall, where a pale stone shaped by Lutyens is the grave marker for all. It was Fabian Ware who calculated that if the British and Empire dead of the First World War were somehow to rise and march, four abreast down Whitehall, it would take them three and a half days to pass that stone. The Unknown Warrior in Westminster Abbey, buried among kings. Anyone in search of that war might choose one of those or some other: the Old Vicarage, Grantches-ter, maybe, or Byron's Pool on the River Cam.

In the grey tumult of these after years
Oft silence falls . . .
And less-than-echoes of remembered tears
Hush all the loud confusion of the heart . . .

So a poor ghost, beside his misty streams,
Is haunted by strange doubts, evasive dreams . . .
And light on waving grass, he knows not when,
And feet that ran, but where, he cannot tell.

5

PENDLE HILL AND WITCHES, LANCASHIRE

'As we travelled, we came near a very great hill, called Pendle Hill, and I was moved of the Lord to go up . . .'

GEORGE FOX, FOUNDER OF THE SOCIETY OF FRIENDS

'If you can see the top of Pendle Hill, it's about to rain; if you can't, it's already started.'

ANONYMOUS

WITHIN THE THREE HUNDRED-ODD SQUARE MILES OF THE FOREST OF BOWLAND, in Lancashire, Pendle Hill looms large. 'Forest' is an echo of old days when the word described a royal hunting ground. Its oldest Latin form – *foris* – means 'outside', 'beyond the tamed' . . . 'other'. People are made other too. Pendle is composed of two words for a hill, being the Cumbric 'pen' and Old English 'hyll'. Mesolithic hunters, knappers of sharp edges left behind by accident or design, knew it as a high top from which to survey the lie of their demesne in the languid millennia after the Ice Age. Bronze Age farmers used the plateau as a burial ground. A rickle of rocks, laid low by time and curiosity and near lost among the mossy mulch, is called the Devil's Apronful after a folk tale about the son of the morning and a bid to level Clitheroe Castle by flinging boulders.

At Pendle Hill there are many stories of many peoples – of hunters, of first farmers, of Romans, of Angles, Saxons and Vikings, come one and all. For the longest time Bowland was woodland wild, home to wolf and boar. In 1652 an itinerant preacher from Leicestershire, George Fox, sought the summit, 'with difficulty', and had a vision: 'From the top of this hill the Lord let me see in what places he had a great people to be gathered.'

On the stumble down, he refreshed himself at a spring that bears his name, then made for the nearby village of Downham, and the inn, where the keeper became the first convert to what would be the Society of Friends, the Quaker movement, whose followers tremble at the word of God. One hill, God and the Devil both.

From Barley, on the Pendle side, the hill looks barren. On the other, where the Ribble babbles, the view is soft, a reminder that it depends where you stand, what perspective you have on the truth of a matter.

In the scheme of things Pendle is not much of a hill – just eighteen hundred feet or so tall and a couple of miles across – but since it stands aloof from the spine of the Pennines, it has the place to itself and seems more. Climb the hard yards to the spot on the Ordnance Survey map called Big End, and the summit beyond is bleak moorland, souls departed. The sea is in sight, if the weather is kind, also the settlements of Burnley, Clitheroe, Colne, Padiham. The flattened top is Pendle Grit, a variation on the theme of carboniferous sandstone spread over limestone, all of it laid down between 290 and 350 million years ago.

In 2011 engineers building a safety wall for Lower Black Moss Reservoir, to hold back the water's dark weight, found the shell of a seventeenth-century cottage beneath a grassy mound. Within the fabric of one wall, they found entombed the bones of a cat. Word soon spread that the place had been home to a witch, and for good reason.

At the beginning of the seventeenth century a movement was abroad to enclose the land, not just in Lancashire but elsewhere. What had been open to all was bounded for the few, embanked and walled and fenced. Vested interests were centralizing wealth, as they always do. The poor were left on the outside in yet another way. Good Queen Bess was a decade dead. James I of England, VI of Scotland, had supplanted her on the throne since 1603. Born and christened Catholic (the priest would have spat into the babe's mouth, as custom dictated, but his mother, Mary, Queen of Scots, had said no), he had been raised a Protestant. Out in the wilds, in Lancashire for instance, there were those who kept to older papist ways, recusants, refusing to accept the Protestant faith (in their hearts, if not in public), and in forgotten valleys far from London, rebellious priests would hear their whispered confessions.

The foundations of Whalley Abbey, sometime home to Cistercians, had been laid at the end of the thirteenth century. By the sixteenth its cloisters were as rank with corruption as any other, but when Henry VIII dissolved it with the rest, its sudden absence hit the locals hard. Children miss bad parents just the same as good. The folk around Pendle Hill were exploited by the fathers – but they were their fathers. Lancashire and the Forest of Bowland, was strongly Catholic. Reformation, dissolution, the ripping away of Whalley's influence had left a vacuum, a moral vacuum too, and into the void had come witchcraft.

Among the poor there were none more vulnerable than women without husbands – widows or spinsters. It was all too easy for such to slip between the cracks and fall ever further. Those who could might earn a few coins as healers, keepers of ancient natural wisdom. If they enjoyed some small success in plying their trade, clearing warts, helping young mothers tend their infants, their reputations might make it worth being known as witches. If they were feared, muttering spells and curses as they grew older in half-remembered Latin or other doggerel, so much the better. It kept others at arm's length and so trouble at bay.

Around Pendle Hill, in a territory beset by lawlessness and licentiousness, where all manner of superstition and old ways had taken root in corners out of sight, more than one family eked a living out of witchcraft. There came a day in 1612 when one Alizon Device (a bewitching name if ever there was one) passed a pedlar from Halifax, one John Law, accompanied by his son Abraham, on the high road through her patch, and asked him for some metal pins. Maybe she meant to pay for them – hard-to-find items, expensive, associated with spells for love and also for the treatments of warts – and maybe not, but Law refused. By his account, she spat a curse after him. He took no more than a few paces before he collapsed, paralysed down one side of his body. For her own part, Alizon was convinced of her powers and accepted at once that her spiked words had caused the hurt. Contrite, she sought Law's bedside where she begged his forgiveness. Her actions in the aftermath of the event persuaded others of her guilt as a witch . . . so that all of what happened happened.

Most importantly, Alizon Device was from a Pendle family of self-professed witches. Taken before the local justice, Robert Nowell, she confessed that she had had the Devil lame John Law. Questioned on the details and her motivation, she gave up members of her family. Her mother, Elizabeth, was a witch too, she said. Her brother, James, was labelled as well, somehow, and both were brought before Nowell. Who knows why but Alizon said next that she had sold her soul to Lucifer. Carried away in the moment, James said his sister had once bewitched a child. Mother Elizabeth had something else to offer: her own mother, known locally as Old Demdike and a self-proclaimed witch of five decades, she bore a mark left by the Devil after he had sucked her blood.

Alizon's blood was evidently up, and fear, and adrenalin. Next, she accused a neighbouring family, the Chattox Clan, of devilment. It may well be that rivalry was at play. Both families had hard-won local reputations for skill with witchcraft and Alizon, blinded by the madness of it all, might have sought to exploit the situation to take down business rivals. There was bad blood anyway, since the Chattoxes had previously broken into the home of the Devices, a property called Malkin Tower, and stolen all manner of goods. Curiouser and curiouser, Alizon alleged old crimes. She said Anne Chattox, matriarch of the clan, had used witchcraft a decade before to murder four men, including John Device, her father. Chattox and Old Demdike were in their eighties. Both were carted to gaol and confessed to selling their souls to the Devil.

Already four were charged – Alizon Device, Old Demdike, Anne Chattox and her daughter, Anne Redfern, accused by Demdike of making clay figurines for some or other skulduggery. Within days of those charges there was a gathering of Demdikes at the family headquarters, Malkin Tower, where the flesh of a stolen sheep was cooked and consumed. In the heat of it all, Justice Nowell investigated the event and, in short order, eight more were charged with witchcraft: Elizabeth and James Device, John and Jane Bulcock, Alice Grey, Katherine Hewitt, Alice Nutter and Jennet Preston.

It was a kind of madness. Lancashire was riddled with poverty, but no more than other counties. The timing of the Pendle Witch Trials might have had something to do with their intensity – the throne was occupied by a king enthralled by witches. In September 1589, Anne of Denmark should have sailed to Scotland to be with James, her new husband, after a marriage by proxy. Storms en route forced her ship to divert to Norway, badly damaged. A second attempt was similarly beset, and Anne decided to stay put until the spring. Frustrated by the delay, James sailed to Norway and spent months there with his bride, then in Denmark, bringing her to Scotland in 1590. Witches were blamed for the storms, with several women tried in Denmark and two burned at the stake.

In Scotland much attention fell upon a healer from Haddington, East Lothian, named Agnes Sampson, accused of witchcraft by one Geillis Duncan (a name familiar to those who have encountered the *Outlander* novels of Diana Gabaldon). Duncan was a healer too, though younger than Sampson, and made her claims about her elder after she had been accused of witchcraft and confessed after torture with thumbscrews.

It was widely believed the Devil was in the habit of licking his would-be servants 'in some private part', an act that left a permanent mark. That most of those accused of witchcraft were women makes even more sense when the sexually abusive nature of the interrogations is considered. In the case of Sampson, she had, according to the official court report, 'all her hair shaven off, in each part of her body, and her head thrawen [tightly bound with a rope] ... being a pain most grievous, which she continued almost an hour, during which time she would not confess any thing until the Devil's mark was found on her privates, then she immediately confessed whatsoever was demanded of her ...'

Even King James – who would later write his own treatise, *Daemonologie*, warning of the dangers of suffering witches to live – poured scorn upon some of Sampson's fantastical confessions, until apparently she took him aside and whispered in his ear the words he and Anne had exchanged, in Oslo, in the privacy of their bedchamber on their wedding night. All the devils in Hell could not have learned as much, he declared, and changed his mind: he knew old Agnes for a witch right enough. On 28 January 1591 she was taken to Castlehill, on Edinburgh's Royal Mile, and there throttled with the garrotte before being burned to death. Thousands of women and men were accused of witchcraft in Scotland during the sixteenth and seventeenth centuries. It is not clear how many died like Agnes Sampson.

James's *Daemonologie* was published in 1597, and after 1603, and the Union of Crowns, his interest had infected his southern kingdom too. A witch could be known, he had written, by her use of natural remedies: 'By curing the Worme, by stemming of blood, by healing of Horse-crookes, by turning of the riddle, or doing of such like innumerable things by words, without applying anie thing, meete to the part offended, as Mediciners doe.'

By 1612, Pendle was ready for its most hysterical and fevered witch-hunt. While recusants hid their popery, wise women and healers had been in plain sight, vying with one another for the greater share of notoriety that, turned into coins or bartered goods, might keep food on the table, a roof overhead. Although a round dozen were due to stand trial at Lancaster, only eleven made it to court. Old Demdike, in her eighties, did not survive her weeks in the filth of her dungeon and died there in the dark.

A key witness throughout was Jennet Device, just nine years old. Even by the standards of seventeenth-century legal proceedings, that the unqualified testimony of one so young was accepted without question was exceptional – but for witch trials, advised King James, all normal conventions might be set aside. Seized, perhaps, by the drama of the days, Jennet gave damning evidence against her mother, sister and brother. Elizabeth howled and cursed her daughter, and had to be dragged from the dock. For whatever reason, dizzy with the fame perhaps, Alizon Device maintained throughout that she was a witch, and when confronted with the pedlar John Law, fell writhing on the floor of the court, crying and confessing her darkness.

Of the eleven that stood trial, ten were found guilty of murders by means of witchcraft, deaths dating back years. Alice Grey was acquitted and lived to tell the tale. Jennet Preston was tried and convicted at York Assizes and hanged at nearby Knavesmire. Alizon, Elizabeth and James Device, John and Jane Bulcock, Anne Redferne and Anne Whittle, Katherine Hewitt and Alice

The young witch Jennet Device sets her familiar, the black cat Tib, upon a victim – contemporary illustration

Nutter were found guilty and hanged together on Lancaster's Gallows Hill, on 20 August 1612.

A generation later, in March 1634, a woman named Jennet Device was among twenty people tried for witchcraft in Lancaster Assizes. It is not possible to say for certain if it was the return to that court of the star witness of twenty-two years before; if it was, her childhood mischief had caught up with her. Whoever she was, the same or a namesake, she and the rest of the alleged coven were spared the rope, or burning or any other horror. Judging by available records, however, she might have spent the rest of her days behind the bars of Lancaster Gaol. Forty years after those nine were hanged in Lancaster, the founding father of the Quaker movement had his ascent and his vision, a message from a nearby God. No one has reported any more recent sightings of Satan on the hill, so presumably the fallen angel, at least, has abandoned the place for more seemly circumstances.

Pendle Hill is famous for its ghosts, most of them supposed to be those of the witches, as though in death they were drawn back, fallen leaves, by its dark mass, its gravity. Others have died on the hill over the years, including suicides, so the landscape may be imbued with all manner of misery. There are reports, too, of the spirits of airmen who came to grief on one or other of the hill's dark faces.

Sadness is the abiding emotion encountered in the shadows there, in the presence of figures half glimpsed, drifting out of sight round twists on winding tracks, flitting through trees or by the gable ends of buildings. A few have sensed malevolence in the air, or anger unspent and simmering that had them pick up their pace down steep slopes in search of easier company. Some visitors, and some locals, have reported the touch of unseen hands, tight grips on occasion.

During her testimony, young Jennet said her mother had kept a brown dog, a familiar, called Ball. More than once, said Jennet, she had seen Ball and her mother talking together, planning murders. Visitors to Pendle Hill have encountered a dark dog from time to time that comes and goes without a sound.

At the Palace of Holyroodhouse, in Edinburgh, at the foot of the Royal Mile, where old Agnes Sampson had met her awful death in the century before, they say her naked, shaven, tortured shade walks the corridors and halls . . .

6

BEN MACDHUI AND THE BIG GREY MAN

'First to our ancestors who lie in barrows or under nameless cairns on heathery hills, or where the seal-swim crashes the island narrows. Or in Jacobean tomb, whose scrolls and skulls carry off death with an elegant inscription. The Latin phrasing that beguiles and dulls. The bitter regret at the loved body's corruption or those who merely share the prayer that is muttered for many sunk together in war's eruption. To all, clay-bound or chalk-bound, stiff or scattered, we leave the values of their periods. The things which seemed to them the things that mattered.'

W. H. AUDEN AND LOUIS MACNEICE,
'THEIR LAST WILL AND TESTAMENT'

I AM HAUNTED BY A THOUSAND THINGS, TEN THOUSAND. I AM HAUNTED BY THOSE lines above, 'the loved body ... things that mattered'. I know them off by heart and quote them, in word or in print, any chance I get. Mostly I am haunted by stories about people, people and their places. I am haunted by a woman and an infant (maybe mother and child, who can say for sure?), buried side by side and together at Vedbaek, in Copenhagen, seven thousand years ago, before Vedbaek was, or Copenhagen, or Denmark. Maybe they died together, those two, in the struggle of birth. She was seemingly buried in an ankle-length dress trimmed with animal bones, a necklace of red deer teeth around her neck. The baby (he or she, too young to tell) was laid upon a swan's wing. Perhaps it was no more than a comforter placed by a husband and father, or was it inspired by thoughts of souls returning in due course in the manner of migratory birds that leave and then come home? I am haunted by the thought that love, and its necessary travelling companions, heartbreak and grief, committed to the ground millennia ago might survive to be unearthed along with stones and bones.

I am haunted by the image I have in my head of a tiny, bird-like youngster, named Birka Girl by a modern world that could never know her real name and called her after the Viking town that went to sleep alongside her, thirteen hundred years ago, on the island of Björkö, in Lake Malaren, in Sweden, making of her a Sleeping Beauty. Her bones, slight as straws, a necklace of glass beads and a gilded brooch are on display in a little case in Stockholm Museum. Beside them stands a life-size realization of what she might have looked like. Since she was buried with wealth, in a place of the highest honour close by Birka's defences, it has been thought she might have been laid down in a red dress – in old days the most expensive colour to obtain. So, a little girl in red. I think about her often, especially when I glimpse a flash of red in my peripheral vision. I am haunted by the thought of those people and more than I could ever list.

HAUNTINGS

Now I am haunted by memories of my dad. He was eighty-seven when he died but I seldom think of him so old. Mostly when he comes to mind he is half that age, slim and straight and handsome. Dad looked young for his years for the longest time. We used to joke about it sometimes, the Peter Pan of Pops.

I don't dream much, or not that I remember on waking, but I have been dreaming about him. Most recently he was there in a moment or two of a dream in which nothing much happened at all. He was with my mum, and some others whose faces I did not notice. Dad was healthy and smiling. He was well turned out too, in smart clothes that looked new. I walked up to him and kissed his cheek and neck. I felt the stubble of his jawline. He whispered something but for the life of me I cannot remember what. I woke up with an unfamiliar feeling of calm.

Haunted . . . haunting: words with roots winding into unexpected places. The Old French *hanter* describes the habit of frequenting a place or visiting regularly. The Old Norse *heimta* is about going home or bringing home. Etymologists have suggested that the idea of a spirit or ghost returning to a place it once knew or to a house in which it had once lived might be Proto-German – but concede that, if so, it has been lost, or even buried, like the dead. Did Rupert Brooke's shade go home? Or those of the Pendle witches? My *Oxford English Reference Dictionary* also offers, by way of a definition of haunting: '. . . of a memory, melody, etc., poignant, wistful, evocative'.

This book is partly about longing. Ours is a species prone to nostalgia, or longing for home. 'Nostalgia' is made of two Greek words – *algos*, meaning 'pain' or 'distress', and *nostos*, meaning 'homecoming'. We humans, some of us, tend to long for that which we have lost or given up, that we know to be out there somewhere out of sight and reach. Sometimes we see what we want to, or need to, whether it is there or not, like the little girl in red.

The dour dome of Ben MacDhui, all 4,295 feet of him, looms above the Lairig Ghru (which means 'the hill overlooking the rivers') in Scotland's eastern Highlands. Both words of Ben MacDhui are Scots Gaelic: 'Ben' means 'hill' and 'MacDhui' might be a corruption of 'MacDhuibh', which is 'MacDuff', so MacDuff's Hill, highest of the Cairngorm range. Otherwise 'Dhuibh', which also means 'black', might refer to the shape of a hog's back, which the Ben resembles from some angles. In all of Britain, only Ben Nevis is taller, by a hundred feet or so. On Ben MacDhui the terrain is of the most rugged sort. The summit and the slopes all about are littered with boulder fields, with cliffs and corries and evidence of Nature at her ceaseless work of transformation. The mountain is being reduced remorselessly by the processes of erosion, ground to dust. Cycles of warming and cooling cause the rock to crack and shatter, audibly if not visibly. The weather feels close there, all-encompassing at times and immanent. As Nan Shepherd writes in *The Living Mountain*,

> Gales crash . . . with the boom of angry seas; one can hear the air shattering itself upon rock. Cloud-bursts batter the earth and roar down the ravines, and thunder reverberates with a prolonged and menacing roll . . . Mankind is sated with noise; but up here, this naked, this elemental savagery, this infinitesimal cross-section of sound from the energies that have been at work for aeons in the universe . . .

Since time out of mind there has been talk around Ben Mac-Dhui of a presence they call, again in Gaelic, Am Fear Liath Mòr, the Big Grey Man. He (and no one doubts it is a he) haunts the Ben's contours. Many are the accounts of moments in the mist and fog when climbers and walkers have heard heavy footsteps following them, crunching in the scree and taking long strides, longer than those of any normal man. There have been glimpses, too, of a thin, looming figure, much taller than a human being, twenty feet tall, some say. Those who have encountered him at close quarters report an all-over covering of hair, in the manner of the North American Bigfoot, or the Yeti of the Himalaya.

Ben MacDhui is, it must be said, a hill enveloped in unease and foreboding as often as in mist. That, in itself, is hardly a phenomenon limited to that mountain. Many and varied are the tales of sensations of unease, sadness, even of panic, in the mountains of the world. Mountain panic is an accepted condition that climbers report, a sudden sense of oppressive gloom, of anxiety and sometimes a need, all at once, to be elsewhere. A feeling that is drifting and amorphous in other high places, however, has a very definite shape and character on Ben MacDhui. In terms of recorded evidence, the story begins in 1891, when a professor and member of the Royal Geographical Society named Professor J. Norman Collie had the first reported encounter with the Big Grey Man. As he said in a speech to an AGM of the Cairngorm Club in 1925, he had summited the Ben alone and was on his way back down when he became aware of the sound of footsteps not his own. As far as he could tell, the other's strides were much longer than his by a factor of three of four . . . The mist was such that he could see little and, although he tried to persuade himself he was giving in

67

to nonsense, he was soon running downhill, for a distance of four or five miles.

'Whatever you make of it, I do not know,' he told his audience. 'But there is something very queer about the top of Ben MacDhui and I will not go back there again.'

It is worth noting that apparently Professor Collie had something of a reputation as a practical joker – but those who knew him well said they believed he was sincere in his account of that place and that day thirty-four years before. He was a respected scientist – professor of organic chemistry at University College London, and a fellow of the Royal Society. Perhaps more relevant, he was an experienced and acclaimed mountain man of the Victorian era: he pioneered climbs on the Isle of Skye and in the Alps. He was part of the team that made the first attempt on the 8,000-metres peak of Nanga Parbat, in the Himalayas, in 1895, and later completed twenty-one first ascents in the Canadian Rockies, including that of Mount Collie, which bears his name. Sgurr Thormaid – Norman's Peak – on Skye is named for him too.

In any event, it was in the years after Collie's claimed encounter that more tales of a Big Grey Man began to accrete around that great dome, shaped like the exposed crown of an all-but-buried head. A recurrent theme is talk of dread on the summit and on the surrounding slopes. Another climber, Richard Frere, described an experience, in the years after the Second World War, involving a presence on the mountain, 'utterly abstract but intensely real'. He also reported hearing 'a high, singing note'. A friend of Frere, who preferred to remain nameless, was camping on the Ben when he awoke, in the night, filled with an intense and ominous foreboding. On opening the flap of his tent and looking out, he saw a tall man, much too tall, with dark hair, silhouetted against a full moon. Whatever it was, it moved away from him and then down the slope with a swaggering gait. On another occasion a writer, Wendy Wood, recalled hearing music as she climbed towards the summit, music of a Gaelic flavour, before hearing footsteps from a source she could not pinpoint, which caused her considerable alarm.

A majority of the encounters with the Big Grey Man are made only of emotions and sounds: sometimes music; the sound of footsteps nearby of a weight and length of stride suggesting someone or something excessively tall and long-limbed; feelings of sadness or impending doom, even a sense of panic and a need to run, hell for leather, towards lower ground. But, like that anonymous friend of Richard Frere, there are those who claim visual sightings too.

The June 1958 issue of the *Scots Magazine* carried naturalist and climber Alexander Tewnion's account of his own grand adventure on the hill. After ten days in the Cairngorms in the autumn of 1943, there came an afternoon when, having summited the Ben in question, he had begun his descent. Dense grey mist originating from the area of the Lairig Ghru rolled in all around him. The very air seemed loaded with ill-will. A freezing wind snatched at his clothes and whinnied among the nearby boulders. Perhaps, he thought, he heard footsteps other than his own . . .

I am not unduly imaginative, but my thought flew instantly to the well-known story of Professor Collie and the Fear Liath Mhor. Then I felt the reassuring weight of the loaded revolver in my pocket. Grasping the butt, I peered about in the mist, here rent and tattered by the eddies of wind. A strange shape loomed up, receded, came charging at me! Without hesitation I whipped out the revolver and fired three times at the figure. When it still came on, I turned and hared down the path, reaching Glen Derry in a time that I have never bettered. You may ask was it really the Fear Liath Mhor? Frankly, I think it was.

A dear friend of mine, an archaeologist, had his own story about the Big Grey Man. Tom Affleck directed the first archaeological dig I took part in as a university student in 1985. Afterwards, and until his early and unexpected death just a few years later, he was a mentor. He had piloted a Spitfire in the last year of the Second World War – which, all by itself, would have made him enthralling. Having first studied for a degree in botany, in the years either side of the war, he had returned to university after early retirement in the late 1970s and completed a second MA, in archaeology. Tom was unforgettable, so my memory of him is haunting.

He and his wife, Isobel, had kept Irish wolfhounds. Tom was tall and lanky, with a tendency to stoop. With his shaggy hair and beard, grey of course, and occasionally lugubrious air, he had something of the hangdog, hound-dog look himself. Or a Grey Man. He was all about stories – about the war and much else. In any gathering he would hold court. More than once he told me about a particular day's hillwalking in the Cairngorms.

He had reached the bottom of Ben MacDhui alone, an hour or so ahead of his climbing partner (who had made a detour to take a photograph of some view). It was a fine day, clear for miles, and Tom described how he had turned and sat down to drink from a flask of coffee and watch for his companion's approach. In due course he saw him in the distance and, striding behind him, looming over him, a grey figure of twice his height. They disappeared together into dead ground for a few minutes – the friend and Am Fear Liath Mòr – and when the friend reappeared, he was alone again. As soon as they were reunited, Tom regaled the other with an account of what he had seen. The friend had been none the wiser, had seen and heard nothing. Tom was a storyteller, for sure, but he was ready to swear the truth of what he had seen that day on the foothills of the Ben. Whenever he told his tale, and I would ask him the hows and whys, he would only shrug.

'It's all I can tell you,' he said once, in answer to my questions. 'That's what I saw.'

If Tom and the others – Collie, Frere, Wood, Tewnion *et al* – had no rational explanation for what they had seen, or felt, there are those of a scientific bent who believe they have something to offer. For one thing, Am Fear Liath Mòr is often explained as a manifestation of a well-known natural phenomenon called a 'Brocken Spectre'. Named for the Brocken Mountain, in the Harz range in northern Germany, where sightings of the effect are common, a Brocken Spectre is the often-magnified shadow of an observer cast against low cloud or mist by diffused, low-angled sunlight directly behind. When the cloud or mist is especially heavy with tiny water droplets, the 'spectre' is enhanced by a rainbow-coloured corona around the shadow. The domed shape of Ben MacDhui's summit, slopes parallel to the angle of the sun's rays, may mean walkers at the beginning or end of a day, especially in winter when the sun is low in the sky, are in locations

71

Spirit in the sky . . . a Brocken Spectre encircled by a corona

shaped for encounters not with the Big Grey Man but with their own elongated shadows.

Long before Collie's tale, and the rest, poet James Hogg was minding his sheep on the border hills when he spied something that briefly terrified him: 'It was a giant blackamoor, at least 30 feet high, and equally proportioned, and very near me,' he wrote. 'I was actually struck powerless with astonishment and terror.'

Just as others would do in years to come, Hogg fled the hill. Returning next day to recover his sheep, he had a second encounter with the apparition. Standing his ground this time, he removed his hat . . . and saw the giant do likewise. All at once – and while the scientific explanation might have eluded him – he understood that he was confronted by no more than his own shadow cast against the fog.

HAUNTINGS

If optical science provides an explanation for what some claim to have seen on the Ben, what about the dislocated sounds, the feelings of unease, of sadness and of dread? In 'The Ghost in the Machine', a paper published in the *Journal of the Society for Psychical Research*, authors Vic Tandy and Tony R. Lawrence described a 'haunting' experienced by Tandy and a handful of his colleagues in a factory manufacturing medical equipment. The building comprised two garages, built back-to-back and repurposed to create one long, open-plan space. Workers reported intermittent feelings of unease, sudden chills. More than one felt the invisible presence of another. On one occasion a colleague was sure Tandy was by his side and turned to speak to him ... only to see Tandy many yards away at the other end of the building. One night, working alone in the building, Tandy felt uneasy again while seated at his desk, chilled as well, hackles raised. All at once he was sure he was being watched and slowly a figure, grey and indistinct, seemed to drift into his peripheral vision on his left side. Briefly terrified, he turned his head to confront the sight head on and found the apparition – if apparition it had been – was gone.

The next night, again alone, Tandy set aside some time to fit a grip to a fencing foil he planned to use in an upcoming competition. The necessary equipment was at hand, and he clamped the foil in a vice. Leaving it in place, he sought some oil to lubricate the cutting of a thread into the tang at the blade's end – and on his return found said blade vibrating wildly. Confronted by a physical effect he understood (Tandy was an engineer), he suspected the foil was affected not by a supernatural presence but by a low-frequency sound wave. Long, narrow spaces were, he knew, ideally shaped for the propagation of such natural phenomena.

Since a frequency lower than 20Hz would be all but inaudible – especially in a factory environment in which other sounds would serve to mask it – Tandy experimented by moving the foil, in its vice, up and down the space. Towards the centre of the room, the vibration was most vigorous, a veritable thrashing, like a metal tail. Moved to either end of the room, the vibration reduced, then ceased altogether, enabling Tandy to draw the conclusion that a low-frequency sound wave, generated by some piece of equipment in the factory, was rippling up and down the length of the space. With nowhere to go, rebounding off the flat walls at either end, the sound was creating a standing wave – a static area of vibration in the centre of the room.

The source was subsequently found to be an extractor fan in a cleaning area at one end of the factory. Once its mounting was changed, the effect vanished. Gone, too, were the feelings of anxiety and unease Tandy and his colleagues had experienced. Further research revealed low-frequency sound has measurable physiological effects on people, some being more susceptible than others. Sustained exposure to such vibrations may affect vision (arguably explaining Tandy's 'grey man' apparition that disappeared when he moved his head and so minutely altered his interaction with the wave) while also creating feelings of anxiety, depression, even fear, after prolonged exposure.

The Brocken Spectre is a common sight and almost a tourist attraction in its own right – even if no one there has ever reported the vision rushing towards them, far less the urge to fire live rounds at it from a revolver. I read 'The Ghost in the Machine' and nod at reasoned argument from scientists about waves of inaudible sound rippling the length of narrow rooms where fans and other kit muster their invisible tempos. I remind myself always about Occam's Razor and how two and two make four with none left over. And then I think about the shattered summit and slopes of Ben MacDhui where folklore says a tall grey sentinel stands ready to unnerve the unwary, to chase from his mountain fastness those he does not want. Would the wind ever make low-frequency sound of the right sort to generate disorienting, disabling waves that make a person see things that aren't there, feel fear for no reason? Hemmed in by mists, could such a wave of noise rebound and stand, creating pockets of space where high hopes disappear, replaced by dread?

All around us are phenomena that science has proved to be there but that we cannot usefully detect or identify with our five human senses alone. While we listen to one radio station, countless others are silently present nearby on slightly different frequencies. There, but not there. The very air is thick with packets and streams of digital data on the move that we neither see, nor hear, nor feel. Our human eyes are sensitive only to a narrow spectrum of what we perceive as visible light and colour. Either side of our limitations are entire realms of other light we cannot see, but which are there just the same. Of what else, how much else, are we blithely unaware on account of the limitations of being only human?

Since the 1960s, physicists have considered String Theory, which says everything in the universe is made up (when you get right down to it) of tiny, one-dimensional strings, all shivering and vibrating. The vibration of each, its resonant pattern, determines its nature. Some vibrate themselves into photons, others into particles within the infinitesimal universe that is the nucleus of an atom, and so on *ad infinitum*. One way or another, say those physicists, it is all and only vibration. So many mysteries hard for most of us to grasp. The solids we feel, floors and walls and all the rest, only seem so to our poor senses because their constituent particles are vibrating so fast, stopping us passing through in the same way the blade of a spinning fan appears like a solid disc, even though it is mostly nothing, mostly empty space.

What else is around us unseen and, as yet, undetected? Up high on a mountain like Ben MacDhui, what else is there to be felt that has neither name nor nature that we know of? Nan Shepherd writes, 'I believe that I now understand in some small measure why the Buddhist goes on pilgrimage to a mountain. The journey is itself part of the technique by which the god is sought. It is a journey into Being.'

7

IONA AND A GHOST OF THE PAST

'Heaven and earth are only three feet apart but in thin places the distance is shorter yet.'

CELTIC FOLKLORE

'. . . the insular Ancient Celts never existed.'

SIMON JAMES, *THE ATLANTIC CELTS*

DO YOU BELIEVE IN CELTS? IT HARDLY MATTERS. HERE IN THESE BRITISH ISLANDS, many do, claiming Celtic roots, a Celtic identity. Plenty have the coiling, interwoven artwork of the tribes tattooed into their hides. For them there is no doubt. Apart from anything else, being, or at least feeling, Celtic draws a line, declares a divide between those who are and those who are not – self-othering, if you will. The sense of other extends to nations then – those with ideas of separatism citing ancient Celtic-ness among the grounds for divorce from the so-called United Kingdom. Not for nothing do some Scots, Irish, Welsh and Cornish tell themselves and each other they have roots winding all the way down to a Celtic past while most English do not. An assumed connection to a pre-Anglo-Saxon, pre-Roman, almost pre-lapsarian Celtic past is a passport to a lost world where everyone and everything was . . . better. Many people – perhaps all people – need something to believe in, whether or not that something is visible and tangible in the real world.

Here in our little archipelago the notion of 'thin places' is attributed to the Celts. Other dimensions – home to ancestors, ghosts, gods, whatever – are close by, they said, and separated from our everyday world by boundaries we cannot see and mostly cannot feel. Here and there in the landscape, however, so the theory goes, there are places where the partitions are worn thin from all that rubbing along together. In such spots a person might sense the glow, warm or cold, of somewhere else . . . somewhen else.

Such an old thought, thousands of years old at least. The rational response might be to dismiss the notion out of hand, except twentieth-century science inadvertently overlaps with and echoes what those so-called Celts said. In hopes of making String Theory properly describe the texture of the cosmos entire, from smallest to biggest, physicists make a radical assumption: rather than one universe made of three-dimensional space plus time, there might be nine or ten or twenty-five or an infinity of universes, all tightly curled and folded together, like a scrum of kittens warm and thriving on closeness, each untroubled by the presence of the others. String Theory allows for uncountable universes all in a cosmic knot of unimaginable complexity – a multiverse. Science or mythology: the boundary between the two is, as it turns out, thin.

Whether they existed or not, no one ever called himself a Celt. The word is a corruption of the Greek *keltoi*, meaning 'outsider'. Most peoples, tribes, called themselves something like 'us' or 'the people'. The Welsh call themselves Cymru, which means 'fellow countrymen'. It was the Anglo-Saxons long ago, invaders of British turf, who labelled them Wielisc, which means 'foreigner'. A Scot was a seaborne pirate out of Ireland, dreaded by the Picts among whom they arrived. Picts were labelled thus by Romans, who saw their body paint or tattoos and applied a name that meant 'painted people'. It was and is the same all over: we don't get to choose what people call us. Rather, we are given names by others, and those names stick. Even as individuals we are named by our parents, nicknamed by friends or enemies. Most of us just accept the handle and get on with it. Among historians and archaeologists there is no consensus about the existence of a homogeneous group – whatever they might have called themselves – spread across Europe more than two thousand years ago, sharing a language and a culture, knowing themselves as one. More likely there were many distinct groups, each making their own use of iron for tools and weapons, perhaps finding some common ground in their approaches to art, to the shaping of culture, to language and religion, but never at any point understanding themselves as a unified people under one banner and one name.

Something else shared was the notion of 'thin places', which was felt and noted elsewhere in Europe and the wider world, as well as in the British Isles. As an idea it has lasted longer than the Celts, whoever they were, and no doubt it predated them too.

Among others, the island of Iona feels like a thin place to me, and to others. It is also a place of ghosts – of monks murdered and flayed by Vikings, of kings of Scotland and Norway. Separated from the larger island of Mull, off Scotland's west coast, by a thin mile of sea – the Sound of Mull – Iona is a tiny scrap, just three miles long, a mile and a half wide. The saint called Columba put the place on the map in AD 563. Columba was a pet name, from Colum Cille, the dove of the Church, so he, too, was named not by himself but by others. His back story, like the name his mother gave him, is lost in myth. If any or all of it is true, it hardly reveals a dove – a hawk, rather.

According to the most popular version of events (written a thousand years later so hardly likely to be reliable), he was high-born, a scion of the powerful Irish Uí Niall clan, anglicized as O'Neill. A prolific writer and scribe, Colum Cille copied a book, a psalter belonging to a senior churchman named Finnian. He wanted to keep it for himself, but Finnian protested, saying that since the book was his, so too was the copy. For final judgement the matter came before the local king – Diarmait – who sided with Finnian.

Man of the cloth or not, Colum Cille took up arms against Finnian and provoked a battle at a place called Cúl Dreimhne, in AD 560 or 561. Remembered as the Battle of the Book, it may count as one of the earliest disputes over copyright. In any event, as many as three thousand men died in the fighting, and for his part in provoking so much bloodshed, Colum Cille was allegedly exiled to neighbouring Scotland – tasked, as penance, with converting to Christianity there as many as died at Cúl Dreimhne. Perhaps the story was different once, so that Scotland was a reward instead.

The north and east of Ireland – Antrim now – were, by

Columba's time, already wedded in some way to a swathe of west-ern Scotland. Argyll remembers 'the coast of the Gaels', suggesting it was a place colonized by foreigners from across the sea. United rather than separated by that sea, the two territories made up the kingdom of Dál Riata, which means 'the portion belonging to Riata' (whoever Riata was). It was in that time of union that Columba of the O'Neills departed his homeland, with a currach-full of followers, and came ashore, as though directed by God Himself, on Iona. More likely it was all by prior agreement, friends in high places in Argyll making the island available to one of their own.

Surely, too, Iona was already long held dear as a thin place. Go there now and it is a place apart in every way. Some (most?) of the

'But ere the world come to an end/Iona shall be as it was'

otherworldliness is about the quality of light, air, the texture of rock, the colour of the sea and the sand. Some of the bedrock of Iona is Lewisian Gneiss, three and a half billion years old and among the oldest on Earth. The myth has the exiles come ashore at Port-a-Churaich – the Port of the Currach – on the island's south-east tip. From there Columba climbed to the nearest high ground of Carn ri Eirinn – the Cairn of the Back to Ireland. He wanted to see his homeland but could not, although on a clear day Ireland is visible from there. So many place names on Iona are hauntingly lovely. From Carn ri Eirinn a visible curl of pale sand nearby is Camus Cúl an t-Saimh, the Bay at the Back of the Ocean.

I am haunted by Iona. So small, easily circumnavigated on foot in a day, but enveloping and all-encompassing. A dislocated piece of a world elsewhere and of another era. The often azure of the shallows, sand like powdered diamonds, grants the island a Mediterranean hue. I have wondered if, in some undocumented time in his life, Columba knew the brightness of the classical world. Did Iona make him think he had stumbled upon a fragment of the place that had given birth to his Christ?

Long before his coming, with his fierce faith, Stone Age hunters had found the island sometime after the last Ice Age. Archaeologists have unearthed the flinty, fire-scorched traces that speak of those other pilgrims seven thousand years ago.

In their own time Columba and his men – zealots? Fundamentalists? Anyway, free of doubt – set about the business of building a haven and a monastery. They were on the outer edge of a vortex that had its heart in the heat of the eastern Mediterranean, the afterburn of Christ himself. There, other men – Church fathers – had abandoned the madding crowd and sought isolation in the desert. Columba and his sort were more of the same, taking their faith to Ultima Thule, the ends of the Earth, where the sea might be interpreted as a desert of another sort. They wore cloaks of wool they had made themselves. To stand apart they shaved their heads from crown to forehead and built a place of turf, timber and daub. It was monastery and university combined. They grew whatever would grow, harvesting seals and fish from the sea. They prayed and preached and kindled a light. Christians were drawn from all over and Christianity flowed outwards in return, into the godless dark of the heathen – the men and women of the heaths and heather.

Columba was revered by those drawn to him, as one in touch with the invisible. There in that thin place it seemed he could see and hear what others could not. Cnoc nan Aingeal – the Hill of the Angels – is a name that memorializes how a novice watched his master, from a distance, in conversation with entities unseen. He lived and died and was buried. The way of life he planted had firm roots by then and flourished. While elsewhere in Europe Christianity waned, laid low by backsliders seduced once more by pagan ways (pagan from the Latin *paganus*, 'the rustics of the villages'), in distant fastnesses, like Iona, it clung and waited.

Forgotten now by most, Iona was one of two sources of some of the finest works of art in the whole world. Lindisfarne, the holy island in the east of England, was the other. Both took part in the making of the books known as the Lindisfarne Gospels and the Book of Kells. No one knows now which came from which island. Nonetheless Iona shares the glory of illuminated manuscripts that lit up a world otherwise darkened.

Columba had come in the century after Rome had departed

Britain, back into the east. As the decadent old whore retreated, so the wild returned. Elsewhere in the abandoned land an Anglo-Saxon poet (translated by Michael Alexander) considered the remains of Roman ways and lamented in 'The Wanderer':

> A wise man may grasp how ghastly it shall be
> When all this world's wealth standeth waste
> Even as now, in many places over the earth
> Walls stand wind beaten
> Heavy with hoar frost, ruined habitations . . .
> The maker of men has so marred this dwelling
> That human laughter is not heard about it
> And idle stand these old giant works.

After Rome, northern Europe, almost the whole of it, lay fallow and forgotten by the world that mattered. On Iona, and other scraps, hopes and thoughts of different, brighter days prevailed. The Christianity Iona coddled would, in time, spread back into the mainland and see a renewal of that faith. Without Iona, and a few other refuges besides, the world might have taken another path and been altogether different.

Vikings came ashore on Iona in AD 806, drawn by the light and by the gold and silver on the altar and wrapped round holy books. They slaughtered and skinned sixty-eight monks on a beach known now as Martyrs' Bay. The white sand is a playground for families but a thousand years ago it was a butcher's yard where ravens pecked raw flesh. The origin of the name is not quite clear. While the Gaelic, Port nam Mairtear, seems to remember the mass murder, it might be older than the crime. Mairtear may be a corruption of *matra* – which means 'relics' – and might refer to Iona as the burial place of kings.

From the twelfth century onwards, buildings of stone replaced whatever timber and turf Columba and his followers had made. The abbey there now is twentieth century, raised on the ruins of a stone cathedral begun in the twelfth that lasted until the reckoning and wrecking of the Reformation. It was brought back to life between the wars by a churchman, of Fuinary, called George Macleod. He would stand on the pier at Fionnphort, on Mull's furthest west point, look across the sound to Iona and sacred ruins and say he felt as he had as a soldier, moving up to the front at Ypres. Iona summons deep emotions, like fish from deep water.

Before Macleod, Iona summoned Dr Samuel Johnson. His companion James Boswell wrote:

At last we came to Iona but found no convenience for landing. Our boat could not be forced very near the dry ground, and our Highlanders carried us over the water.

We were now treading that illustrious Island, which was once the luminary of the Caledonian regions, whence savage clans and roving barbarians derived the benefits of knowledge, and the blessings of religion.

To abstract the mind from all local emotion would be impossible, if it were endeavoured, and would be foolish, if it were possible. Whatever withdraws us from the power of our senses; whatever makes the past, the distant, or the future predominate over the present, advances us in the dignity of thinking beings. Far from me and from my friends, be such frigid philosophy as may conduct us indifferent and unmoved over any ground which has been dignified by wisdom, bravery, or virtue. That man is

little to be envied, whose patriotism would not gain force upon the plain of Marathon, or whose piety would not grow warmer among the ruins of Iona.

Maybe the fabric of Iona has been worn thin by a river of people. Sands through an hourglass, time after time, make the narrows wider. The friction of grains makes the glass thinner so that what had taken an hour takes less, and time, the fourth dimension, is altered too.

Close by the rebuilt abbey is Torr an Aba – the Hill of the Abbot. It amounts to little: a ridge of higher ground, where dark rock pokes like a knee through worn trousers. It has been called the Hill of the Abbot for a stubborn belief that Columba had a sanctuary there, a little cell he built for himself of wood and turf. It was a place where he could be alone with his God. In the late 1950s Torr an Aba was excavated. Among other bits and pieces, burned wood was found, and although it was in the days before radiocarbon dating, the archaeologist Charles Thomas had the wit to keep the fragments in a matchbox. In 2017, long after his death, other archaeologists came for the box and dated the radioactive carbon in the hazel wood. Columba died on Iona in AD 597 and the radiocarbon dates made plain that the structure on the ridge had been built of timber harvested between AD 540 and 650. There had been a wooden cell there in Columba's day, right enough.

The opportunities to stand in a place where, there is good reason to believe, a named figure from a millennium and a half ago once stood, are rare. Columba – one of those few known for ever by a single name – is almost a legend. And yet on Torr an

Aba, a place remembered by association, the literal truth of him seems within reach. He was on that spot in his time and we can occupy the same in ours, donning his space, like a shed skin.

Near the abbey is the oldest stone building still standing. Called Reilig Òdhrain – the Relics of Oran – it recalls the name of one of Columba's followers, supposedly the first to die and be buried. Reilig Òdhrain is also the name of the graveyard. Legend has it that forty-eight Scottish kings lie buried there – along with others from Norway, Ireland and France – but most historians have doubts. Some royal remains likely lie in that hallowed ground but the inflated number may have been propaganda to embellish the reputation of Iona as a light in a dark world. Macbeth and Duncan may linger, real bones of real men and nothing to do with the avatars imagined by Shakespeare.

Whatever the truth of Reilig Òdhrain, the sentiment is in 'The Iona Boat Song':

Softly glide we along,
Softly chant we our song,
For a king who to resting is come;
Oh, beloved and best,
Thou art fairing out west,
To the dear isle Iona, thy home . . .

In that thin place to the west of the west, shades of monks have been a staple. Martyrs' Bay is haunted by those butchered by Norsemen. Dark, like crows, pale faces and questioning eyes. Witnesses testify to sadness – fleeting, or that stays for days. From time to time the spirits of kings drift noiselessly along Sraid nam

Marbh – the Street of the Dead – retracing their steps along a cobbled way leading from Martyrs' Bay, where their stately coffins were landed after crossing from Mull, for onward passage on the shoulders of others to Oran's graveyard.

In their time, the monks had raised 360 crosses of stone, the empty sort called Celtic, with circles supporting heavy stone arms that might otherwise break under their own weight. No dead Christ upon them, himself risen already. Almost all were smashed by the reformers, thrown into the sea. A few remain.

The haunting of Iona is not about Martyrs' Bay or even the dove of the Church in his cell of timber and turf long ago. It is about something else, something sensed, or someone else close by. You might say it is all ghosts on Iona, souls caught like tufts of sheep's fleece on a rusting nail. Iona is haunting and haunted because the spirit lingering is the peace and ghost of other days. This poem is attributed to Columba:

In Iona of my heart, Iona of my love,
Instead of monks' voices shall be the lowing of cattle,
But ere the world come to an end
Iona shall be as it was.

8

WINDSOR CASTLE, HENRY VIII AND HERNE THE HUNTER

'Therefore, send not to know
For whom the bell tolls,
It tolls for thee.'

JOHN DONNE, 'FOR WHOM THE BELL TOLLS'

ON 9 SEPTEMBER 2022, THE DAY AFTER ELIZABETH II DIED AT BALMORAL, A BELL in the Round Tower of Windsor Castle tolled once for each of her ninety-six years. It was the Sebastopol Bell. One of a pair captured from the Church of the Twelve Apostles, in Crimea, in 1856, it is called into service only to mark the deaths of senior royals, telling their stories and tolling their time. It had not sounded for twenty years, since the death of the Queen Mother.

Everything about the Sebastopol Bell is grim, melancholy. The Crimean War pitted Russia against the combined might of Britain, France, Piedmont-Sardinia and the Ottoman Empire. The city of Sebastopol was besieged in 1854 by Anglo-French troops and held out for most of a year. By a miserable coincidence, the city fell in 1855 on the same date as the bell rang for Her late Majesty's death, 9 September. It was a smoking ruin. The bells were looted as trophies and brought to England. One was eventually installed at Windsor Castle, with a huge Russian cannon seized at the same time; the other made its way eventually to the home of the British Army, at Aldershot Barracks, in north Hampshire. The sonorous notes of the Sebastopol Bell mean death, and only death.

Windsor Castle has a history dating all the way back to the eleventh century and the reign of William I. With a view to defending the western approach to London, the Conqueror chose the site himself. Overlooking the Thames, on the edge of what had been a hunting ground for the Saxons he had usurped, he set his builders to work in 1070, creating, first, a timber fortress. A hundred years later, William's Norman keep was replaced with the Round Tower. The castle had always been primarily a military fortification – first of timber and then of stone but modified only with defence in mind – but Edward III transformed it. From the 1350s until his death in 1377 he spent the better part of fifty thousand pounds – the most invested by any medieval king on a single building – creating a Gothic palace, combining official and private apartments for himself and his queen, Philippa of Hainault.

When Elizabeth Tudor came to the throne, the castle needed root-and-branch repair. After her efforts came those of the Stuarts, some of whom undertook the remodelling of the private apartments among other touches. George III and George IV made their own impact – that of the latter so substantial that little needed doing by his successors. Queen Victoria added a private chapel and had the grand staircase rebuilt. Fire took hold in Victoria's chapel on the night of 20 November 1992 and gutted the space, along with St George's Hall, the Grand Reception Room, the State Dining Room and the Crimson Drawing Room. Elizabeth II counted the blaze as part of her 'annus horribilis'. The necessary repairs took five years.

In short, for a thousand years, Windsor Castle, one of the oldest buildings in England, has seen them come and seen them go, the Reaper having prowled Windsor Castle as determinedly as any hunter. According to those who know it well, more than a few, royal and commoner, have remained behind after their death. Predatory Henry VIII made his presence felt at Windsor both alive and dead. In life he added the grand gate that bears his name, as well as a timber terrace from which he could watch hunting in the park below.

His first wife was Catherine of Aragon, widow of his elder

Windsor Castle . . . uneasy rest the heads that wear the crowns

brother, Arthur. She gave birth to a stillborn girl in 1510 and then, in 1511, a boy, Henry, who lived not quite two months. Two more baby boys were stillborn, in 1513 and 1515, then in 1516 their luckless, unhappy daughter, Mary. Henry and Catherine are such vivid characters, made almost legend by endless telling and retelling of their story, it can be easy to overlook how relentlessly they were stalked by the special sadness that must level high and low.

Always there were mistresses for Henry, unknown numbers, truth be told. Before there was an Observatory at Greenwich, a hunting lodge stood on the site, a favourite place for him to meet his lovers. Henry and Catherine had a second stillborn daughter in 1518. The following year he had an illegitimate son, Henry Fitzroy, with Bessie Blount, the comely daughter of his loyal servant Sir John Blount. Along the way he certainly bedded Mary Boleyn, one of Catherine's ladies-in-waiting, and may have fathered her two sons. In 1525 he fell for Mary's sister, Anne – but since she would not have him for a lover, he embarked upon what became known as the King's Great Matter, his desperate effort to end his marriage in favour of his latest fancy.

In the end it was the Church of Rome he divorced, establishing the Church of England in the process. Catherine was put away, out of sight if never wholly out of mind. He married Anne Boleyn, of course, and thereby embarked upon the journey that would define him in popular memory – the collecting and discarding of wives. By the end of his life, he had divorced two – Catherine of Aragon and later Anne of Cleves. Jane Seymour died soon after bearing him the son, Edward, he so desperately wanted. Catherine Parr, his last wife, outlived the old monster. The other two – Anne Boleyn and Catherine Howard – he condemned to the block.

So much ill-directed heat and also withering coldness. Henry VIII was an emotional enigma. In an interview for the website accompanying his TV series *The Six Wives of Henry VIII*, historian David Starkey describes his behaviour towards women as 'eccentric', even by the imbalanced standards of the day, when infidelity was expected of kings but punishable by death in their queens.

'What is extraordinary is that Henry was usually a very good husband,' said Starkey. 'And he liked women – that's why he married so many of them! He was very tender to them . . . he was very generous . . . He was immensely considerate when they were pregnant.

'But once he had fallen out of love . . . he just cut them off. He just withdrew. He abandoned them. They didn't even know he had left them . . .'

If emotions felt once in a place might leave a residue – like smoke from an extinguished fire – that would explain why Henry's shade, wretched and rotten, is one of the most commonly sighted in Windsor Castle. Both he and Jane Seymour are buried in the vault in St George's Chapel.

More strangeness yet might make one wonder less about the haunted nature of that castle: in that space beneath the chapel's floor lie not just a husband and one of his wives, but the remains of Charles I and an infant. As though to pass it all off as unremarkable, a plain dark marble slab bears the names of the mismatched royals below. The vault was meant only as a temporary destination for Henry and Jane. Like Anne, Jane had spurned his advances until he married her. Directly or indirectly her terms sealed the fate of Anne Boleyn – the latter knelt upright before a French swordsman, who beheaded her just eleven days before Henry's third wedding.

Anne was slaughtered in the Tower of London, Catherine

Howard too, though by the axe, not the sword. Both ladies' remains are bundled among those of fifteen enemies of the state, in a hole measuring just six yards by four, beneath the floor of the Royal Chapel of St Peter ad Vincula inside that tower.

Death hung around Henry like a bad smell and Jane, too, died just seventeen months later, in October 1537. A tomb fit for a king was under construction so her remains were stowed in the vault. Henry followed her in 1547, and while his will made plain that his children were to finish his tomb and move him and Jane into those more fitting circumstances, they were left where they lay, forgotten beneath the floor, like lost heirlooms.

Their time alone was interrupted 102 years later when Oliver Cromwell's Parliament, moving to appease the mob made livid by their regicide, interred the butchered king with his Tudor predecessors. Now there were three swept beneath the floor. How the mighty are fallen.

The world moved on until 1813 when, during the building of a mausoleum at the chapel, during the reign of George III, the forgotten space beneath was disturbed once more. There were found Henry and Jane and, beside them, a skeleton separated from its skull, also the bones of a baby. The latter was Queen Anne's, one of her multiple stillborn, and while a casual observer might have thought the headless corpse belonged either to Anne Boleyn or Catherine Howard, those better informed knew them for what they were: King Charles I. The four were left in place, and eventually William IV commissioned their marble lid.

So much death – as a commoner might expect of a place roamed by vengeful royals for a thousand years. So much horror, too, and awful human sadness. Henry is regularly sighted, wandering the hallways and cloisters. Others claim only to have heard his footsteps – identifiable by the distinctive sound of him dragging one leg.

He died a bloated, rotting wreck, but he had been an athlete in his youth. Fond of the tournament he had been injured in one leg while jousting and paid the price ever after. Subsequent hurts to the leg resulted in a wound that would not heal, remaining a foul-smelling ulcerated mess for years until his death. Some historians have suggested the incapacity that plagued him after those jousting accidents contributed to nothing less than a morbid change of personality, from tender lover into selfish brute. His ghost is said to scream and moan, the agonized wailing of a spirit rightly tormented by pain – pain of injuries and physical corruption, and the pain he inflicted upon so many others.

Five men were sucked into the vortex of Anne Boleyn's downfall, including her brother, George. All were charged with adulterous – in George's case incestuously adulterous – affairs with the queen. Two days before Anne's death, they were executed.

Catherine Howard's despatch was foreshadowed by those of her courtier Thomas Culpeper and her former fiancé, Francis Dereham, whose assignations with Her Majesty, real or imagined, sealed their doom.

If Henry haunts Windsor, his burden of guilt is great and undeniable, deep and sour enough long to outlive him and hang around the hallways. His is hardly the only ghost reported there: Anne Boleyn's has been sighted countless times, roaming the grounds, glimpsed broken and sad at the window of the Dean's Cloister. More than once she has been seen inside a carriage pulled by headless horses guided by a headless driver. Those who have encountered the vision say they know it for Anne Boleyn because the passenger within sits cradling its own severed head.

Windsor – or, at least, the surrounding Great Park – is haunted by Herne the Hunter, a near-mythological figure with many versions of his story. One account has it that in life Herne was a keeper at the Forest of Windsor during the reign of Richard II. Favoured by the king for his superior hunting skills, he inspired bitter envy among his fellows. Framed by them for some crime he had not committed, he was cast out by Richard and hanged himself from an oak tree. The tree in question, near Frogmore House, was the scene of innumerable sightings of Herne's spirit in the form of a figure cloaked in darkness and with stag's horns. In an attempt to exorcize the ghost, Victoria ordered the oak felled and burned, but Herne has never gone away. Any sighting of him is said to be a portent of doom, and of imminent bad luck for the Royal Family.

Shakespeare invoked the haunting hunter in *The Merry Wives of Windsor*:

There is an old tale goes, that Herne the Hunter
(Sometime a keeper here in Windsor Forest)
Doth all the winter-time, at still midnight,
Walk round about an oak, with great ragg'd horns;
And there he blasts the tree, and takes the cattle,
And makes milch-kine yield blood, and shakes a chain
In a most hideous and dreadful manner.
You have heard of such a spirit, and well you know
The superstitious idle-headed eld
Receiv'd, and did deliver to our age
This tale of Herne the Hunter for a truth.

Interestingly – given Henry's serial unhappiness as a husband, not to mention the horrors he inflicted upon some of his wives – another version of the legend makes Herne one Richard Horne, a yeoman in Henry's time who was executed for poaching. Better yet, the image of the horned man is suggestive of the cuckold. A man whose wife is taken by another is ridiculed by having others make the sign of horns behind his head, symbolic of genitals – not his own but those of whoever has replaced him in the marriage bed. Always, the cuckold is the last to know of his wife's betrayal, so the horned figure roaming the grounds of Windsor Castle may be another manifestation of Henry, his accompanying hurt and paranoid bitterness bringing bad luck for his descendants in his wake.

On and on runs the list of ghosts there – reported not just by staff and visitors but also by members of the Royal Family down the ages. The Deanery is haunted, according to some or other of their majesties, by a lad who cries out, 'I don't want to go riding today!'

Bells toll by themselves in the Curfew Tower – sometimes soon after footsteps have been heard on the tower stairs. A wraith identified by some as all that remains of a soldier of the English Civil War, a prisoner, is seen in the basement of the tower, for long used as a dungeon, or, more uncomfortably, experienced as an unseen presence brushing past the shoulders of the living. The ghost of a suicide – a guardsman who shot himself on duty – roams the Long Walk. Before taking his life, he had reported a ghostly encounter of his own. Whenever his apparition appears, witnesses report a smile on its face.

George III, famed for his madness, spent long periods confined to his bedroom, and since his death there have been numerous sightings of his likeness peering out of the window, often in the direction of the guards on duty below. Elizabeth I roams the castle. Among others, members of the Royal Family have, on many occasions, reported seeing her dressed all in black and walking through the library, her high heels tap-tap-tapping on the floor as she makes her way through to an adjoining room.

At least one of the bedrooms in the private apartments is haunted by the ghost of George Villiers, Earl of Buckingham, an especially pretty young man and a favourite of James VI and I. He was one of many who caught the king's eye and the two were rumoured to be lovers. Arguably no other companion pleased His Majesty more, to such an extent that he felt the need to explain his affection to the Privy Council: 'You may be sure that I love the Earl of Buckingham more than anyone else, and more than you who are here assembled. I wish to speak in my own behalf and not to have it thought to be a defect.'

That most devout of kings, who commissioned the translation of the Bible that bears his name, went so far as to invoke his saviour in order to explain the depth of his feelings: 'For Jesus Christ did the same,' he said, 'and therefore I cannot be blamed. Christ had John, and I have George.'

The admiration of the monarch is a curse as often as a blessing, provoking as it does the jealousy and murderous loathing of rivals. Much as Herne was turned on by his fellow hunters, those looking on at Villiers and their king found violence in their hearts. The duke's popularity at court survived the death of James and was sustained by his son and successor, Charles I. Although no one suspected a love affair between them, Villiers was still favoured with elevated position and one promotion after another. Serving Charles as lord admiral – and de facto foreign secretary – his efforts were invariably disastrous. After a string of military failures abroad Villiers was stabbed to death in the Greyhound inn at Portsmouth by an army officer, John Felton, who had seen the other promoted above him once too often.

Poor Charles is among the cast of ghostly characters too. Down through the centuries since his death, one royal after another has claimed to see him nearby, always sad, always forlorn.

Windsor Castle is a monument to many things, not least the longevity of monarchy in these islands. Cromwell instigated an Interregnum, an interruption, but the return of royalty proved irresistible. Windsor has stood by throughout, marking time, with and without the Sebastapol Bell. The place is also a reminder that kings and queens must die like all the rest – and some of their deaths have been as ignominious as any other. Therefore, send not to know for whom the bell tolls . . .

If Dad had a love other than us, it was getting into the car and heading for the hills. The family home is in Dumfries, in Scotland's south-west, and all the places he cared to visit, and revisit, were in the north. Years ago, as part of presenting a television series called *The Face of Britain*, about the genetic inheritance of the

British people, I had my DNA tested. That which I had inherited from Mum, coiled in my mitochondria, showed that her ancestry had been rooted in the west of Scotland for many thousands of years. Like a contented hobbit she was from a long line of people disinclined to leave The Shire. Mum has always been happiest within her own four walls and disinclined to stray. Dad was the opposite, restless, itchy of feet. Of an evening he would sit with maps and an atlas, dreaming of destinations – even if he would only get there in dreams. The DNA I inherited from him had markers, evidence of origins in what was once Persia. It struck me as perfect that while my stay-at-home mum's DNA was where it had always been, Dad's had wandered far and wanted to go further yet.

More than any words of poetry I heard him quote – and he had a decent repertoire of lines learned long ago by rote – he returned time and again to

Up the river and over the lea,
That's the way for Billy and me . . .

I would ask him for more of the words, but if he had ever known them, they were forgotten now. I would ask him too what the poem was called, and he could not remember. It was only after he had died that I bothered to track it down and found the lines belonged to 'A Boy's Song', by James Hogg, the Ettrick Shepherd. It ends:

But this I know, I love to play
Through the meadow, among the hay;
Up the water and over the lea,
That's the way for Billy and me.

I did my best to read the whole of it, all six stanzas, as part of my eulogy for him at his funeral. It was during lockdown when funerals were for immediate family only. My wife and our three children had to sit apart from my mum and sister. My in-laws were there too, also in a row of their own. And that was it – nine of us in a crematorium, my dad in a wicker coffin.

I was the one nominated to speak, watching my children cry.

More than anyone I have known, Dad belonged on the road, heading north. When I was eleven or twelve, he took me on the first of several road trips we made together. We packed some stuff on to the back seat of our canary-yellow Volkswagen Beetle (WCS 224K) and set out on a tour of the west coast of Scotland. Loch Lomond, Oban, Fort William, Skye, Plockton, Ullapool, Durness . . . We stopped at Culloden battlefield, and he took a photo of me crouching by the rough-hewn boulder that, according to tradition, marks the mass grave of the men of Clan Cameron who fell among the thousands of Jacobite Highlanders on that awful day in April 1746. Our family tradition has it that my mum is descended from Cameron of Lochiel, first of the chiefs to step up in support of Charles Edward Stuart, the Bonnie Prince. Mum is Norma Agnes Cameron Neill, her maiden name modified to have a single *l* for mine. In this way are mothers' family names passed on, remembered. I have been back to Culloden many times, back to the gravestone of the Camerons. I try to make it real to myself that some of that battlefield's dead are mine.

9

GLAMIS CASTLE: THE GREY LADY, THE MONSTER AND THE TONGUELESS WOMAN

'I began to consider myself too far from the living and somewhat too near the dead.'

SIR WALTER SCOTT, *LETTERS ON DEMONOLOGY
AND WITCHCRAFT*

O N 17 JULY 1537, A BEAUTIFUL WOMAN WAS LED OUT OF EDINBURGH CASTLE. SHE
was frog-marched across to nearby Castlehill at the top of the
Royal Mile, separated from the fortress itself at that time by a deep
ravine (not bridged until 1753, by the construction of the Esplan-
ade) and there bound to a tall wooden stake. She was the Lady
Janet Douglas, Countess of Glamis, thirty-three years old and
known not only for her long red hair and good looks but also her
compassionate behaviour towards all she encountered. She had,
however, had the gross misfortune to fall foul of her king, James V.

Years before, during his minority, James had been a virtual prisoner of Janet's brother, Archibald Douglas, Earl of Angus. Marriage to James's mother, Margaret Tudor, widow of James IV of Scotland and sister of English Henry VIII, had made Angus the lad's stepfather. The union, always tempestuous and capricious, did not last but after the two separated, the connection inspired Angus to seek control of Scotland by taking control of the boy, in 1526, when he was fourteen. That audacious seizure of custody ended two years later when James was returned to his mother, but the grudge forged then was not forgotten by the young king. As soon as he was able, James had the Scottish Parliament arraign Angus and his closest kin for treason. The errant stepfather dug in behind the walls of his stronghold at Tantallon Castle, by the Firth of Forth. James lost his guns besieging the place, and in 1529 Angus was allowed to flee into exile in England, his estates forfeited, his kin disgraced.

Most of those were later let off the hook but James would exact terrible revenge upon a luckless few. Janet's brother-in-law, John, Master of Forbes, was convicted of planning to shoot the king – a charge no one seriously believed – hanged, drawn and quartered on 14 July 1537. Sir James Hamilton of Finnart, damned by association with the Douglas family, would be executed in 1540, on trumped-up charges of planning another assassination attempt on the king.

And then the Lady Janet. Castlehill, where she stood that day, bound to the stake, was and is one of the oldest places of settlement in the city. It has been the site of homes and a school, a place of worship and celebration, the site of a well of clean water. By Janet's time, it was known as a place for the burning of witches.

NEIL OLIVER

Whatever her sins, it was conviction for witchcraft that was her undoing.

James had been after her, off and on, since 1528 and the first year of his personal rule. She was a Douglas, and that was enough. Perhaps her famed beauty, and the thought of ending it, made her even more of a target for a vengeful man. She was charged, that first time, with providing help and support to her Douglas brothers during their defiance of the king. That legal action came to nothing, but in 1532 it was alleged that the death of her first husband – John Lyon, 6th Lord Glamis, in 1528 – had been her doing by poisoning. Again, the charge was dropped, and she married, next, Archibald Campbell of Skipnish. She was Countess of Glamis once more when, on that July day, 1537, James came after her for the third and final time. It was witchcraft that was alleged, that she had sought, by black magic, to poison the king. As might have been expected, given the charges were fantasy, evidence and witnesses had been hard to come by. Undeterred, James had had her family, including her sixteen-year-old son, as well as her servants, stretched on the rack until finally some broken soul spat out the necessary testimony. Quickly tried and convicted she was sentenced to death by burning.

There on Castlehill, with her son made to watch, barrels of pitch were packed tight around and set alight to make a candle of her. Next day, her husband attempted an escape from a cell inside the castle and fell to his death on the rocks below. No more was heard of poor Archibald, or any of that sad fellowship torn apart by James. But in years to come and down to the present day, a host of witnesses back in Glamis would say Lady Janet had come home.

*

A first sight of Glamis Castle, in the county of Angus, some sixty-odd miles north of Edinburgh, might make a person think it ought to be inhabited by a Disney princess. Rosy pink when the sun shines on its sandstone, it is all frou-frou towers and turrets topped with what look like outsize candle snuffers. The gardens and grounds are vast, and impeccably kept. Perhaps most famous now as the birthplace of Queen Elizabeth the Queen Mother, in truth it has the look of a place where Belle and her Beast would seem right at home. But that's just the place as experienced from the outside.

Most of what is there now dates from the seventeenth century, but Glamis has a thousand and more years of history. Shakespeare's Malcolm was murdered there, put to death by Macbeth,

Glamis Castle and a thousand years of history and hauntings

Mac Bheatha, son of life. The real Macbeth had no connection to the place but there is plenty of gloom about Glamis, Scotland's second-oldest inhabited castle. In the time of Macbeth and Malcolm, in the eleventh century and before, there was a royal hunting lodge on the site. The first castle was built in the second half of the fourteenth century and granted by Robert II to Sir John Lyon, 1st Thane of Glamis and husband to that king's daughter Johanna, known as Jean. The 1st Lord Glamis was Patrick Lyon, John's grandson.

After his despatch of Lady Janet, and the death of Archibald, King James took Glamis for his own for a while. The 7th Lord Glamis got it back, and the 9th, Patrick Lyon, created Earl of Kinghorne, remodelled the place into its present form. Glamis was so mistreated by Cromwell's soldiers during the Commonwealth that it was rendered uninhabitable. It was recovered by the 3rd Earl of Kinghorne, another Patrick, who undertook the necessary restorations and commissioned the Baroque Garden. In 1767, the 9th Earl, John Lyon, married Mary Eleanor Bowes – new money, heiress to a coal-mining fortune – and added her name to his own. Elizabeth Bowes-Lyon was born in 1900 and spent many of her childhood years at Glamis. On 26 April 1923 she married the youngest son of King George V – Prince Albert, Duke of York – and put the castle firmly on the map.

But it was Claude Bowes-Lyon, 13th Earl and grandfather of the Queen Mother, who famously said, 'If you could even guess the nature of this castle's secret, you would get down on your knees and thank God it was not yours.'

In truth a whole cast of characters is said to haunt Glamis. More than one apparition was sighted by the young Elizabeth herself, including that of an African servant boy who worked at the castle in the eighteenth century and was badly mistreated. He is most often seen outside the bedroom that was hers, sometimes perched upon a stone seat and sometimes, allegedly, tripping passers-by. Many are the guests who have reported their bed-clothes pulled off them in the night, scattered on the floor, incidents always attributed to the beaten boy.

Another Black servant was singled out for the cruellest mis-treatment of all. During the fifteenth century a group of aristocrats at Glamis are said to have stripped him naked and hunted him through the grounds with their dogs. Finally cornered, the man was torn limb from limb and now his ghost is to be glimpsed run-ning through the hallways, terror in his eyes.

A room known as the Hangman's Chamber is haunted by the spirit of a servant who hanged himself there. Elsewhere there are reports of knocking at doors and on walls, and places where the air turns suddenly and unseasonably cold. There are sightings of a towering, armoured knight who looms over guests wakened in their beds – and also of an old lady, bent with age and carrying a bundle of laundry towards the middle of the main courtyard where she disappears into thin air.

It is tempting, easy, to mock such testimony. Perhaps it matters, though, to acknowledge that certain places in the landscape, thin or not, attract and hold more than their share of stories. It begs the question of why tales of haunting are not evenly spread, like butter over toast, but clumped in the vicinity of particular places.

In 1486, during a time of feuding between two rival clans – the Ogilvies and the Lindsays – a party of Ogilvie men, pursued by an overwhelming force of their enemies, are said to have sought sanctuary inside the castle. Unknown to them, the Lord of Glamis, their erstwhile saviour, was in league with the Lindsays and led the frightened men to a secret, windowless chamber. Once they were all inside, he locked and barred the door against them. When the leader of the Lindsays arrived, the Lord of Glamis could assure him he would have no more trouble from that particular band of Ogilvies. The room remained sealed until the men were long dead, their bodies later disposed of. Visitors shown the room now will often report the oppression of gloomy thoughts, of sadness, or a sense of impending doom.

That secret referred to by the 13th Earl – Claude Bowes-Lyon, grandfather to the Queen Mother – the knowledge of which might put a man on his knees in gratitude that the castle was someone else's burden – likely concerns Glamis's most famous sadness. More than one researcher has concluded that a severely deformed or disabled child, a son and heir, was born to Thomas, Lord Glamis, and his wife, the great-great-great-grandparents of King Charles III. Records show that a son was born to them on 21 October 1821, and that he died the same day. According to the dread secret, though, the boy lived. Since he was deemed 'so unpresentable' he had to be kept forever out of public sight. When the author James Wentworth-Day was researching a history of the Bowes-Lyons, he was a guest at Glamis for some weeks. From some or other family member he heard testimony that the luck-less heir was 'a creature fearful to behold . . . a deformed caricature of humanity . . . his chest an enormous barrel, hairy as a doormat, his head ran straight into his shoulders and his arms and legs were toy like'.

Deformed or not, 'unfinished . . . scarce half made up', like a latter-day Richard III, he had to be cared for and was kept in a suite away from prying eyes. If the legend is true, he was the rightful heir to the title and the lands – so that his wrongful dispossession made liars of all who came after him and knew the truth. However long he lived, whatever his circumstances, he died and was buried, secretly, on the estate, his suite bricked up and forgotten. There are versions of an anecdote that has a party of guests determined to find his apartments going from room to room – all that they could find – and hanging towels in the windows. When they gathered in the courtyard below and looked up, there were several windows in a row without towels and, no matter how hard they searched the interior, no door to be found.

Arguably as upsetting as the thought of the reality of the so-called Monster of Glamis is the legend of the Tongueless Woman, who wanders the halls and grounds with blood pouring from her mutilated mouth. She is seen at windows, too, pointing to her wounds. Taken with the tale of the entombment of the Ogilvies, her contribution seems to reinforce the notion of a secret at Glamis, something buried or walled up, something that must not be spoken of. The Tongueless Woman is supposed to have overheard or otherwise stumbled upon a family secret. Hear no evil, see no evil, speak no evil . . . When she blurted out her intention to reveal what she had learned, she was first silenced, then murdered.

Yet another tale has a Lord Glamis drunk on a Saturday night and insisting on playing cards. Since the Sabbath was nigh, none would take him up on the offer, gambling being a sin. In his rage he bellowed that, Sabbath or not, he would play cards even with the Devil himself. Soon after, a tall stranger, a dark man, arrived at the main door and, having been welcomed inside, offered to play. Lord and stranger retired to a small room and the door slammed behind them. Soon there was much shouting and swearing. When a servant bent to peer through the keyhole, he was blinded in one eye by the brightness within. The dark stranger was the Lord of Light, Satan himself, and, having won the soul of Lord Glamis with the first hand of cards, damned his host to play on until Judgment Day. The chamber in question is said to have been in the castle's west tower, and there, guests have reported hearing the rolling of dice, raised voices and the banging of doors.

But what, after all this time, of poor Lady Janet? Loneliest and saddest of all, with so much to be sad about, her tortured spirit roams the castle's clock tower and, most often, the family chapel. Some sources claim she seeks her tormentor, King James. On one occasion a hundred witnesses all at once saw her apparition moving through the chapel before sitting in one of the seats at the end of a row by a wall. Visitors are invited to try the seat for themselves, but during services it is left empty for the Grey Lady: Lady Janet.

So many stories at Glamis, so much sadness. Might it be permissible to think all of it has gathered not by accident but rather like the dust around the electric charge of some suppressed truth? A hidden chamber where an inconvenient man (or men) died out of sight and out of mind ... A woman silenced ... Lady Janet, innocent, wronged and condemned to that worst of deaths.

Given the feel of the place, its gloomy reputation, the castle was often largely abandoned by family and left in the care of factors. It was in such circumstances that the writer Walter Scott asked to spend a night at Glamis, in 1793. He was shown to his room and the factor withdrew, leaving him to it. Whatever he saw, or felt, he did not write about his stay until nearly forty years later. In *Letters on Demonology and Witchcraft* he noted

> ... that degree of superstitious awe that my countrymen call *eerie* [when spending the night at the notoriously haunted castle] ... I must own, that as I heard door after door shut, after my conductor had retired, I began to consider myself too far from the living and somewhat too near the dead.

Remember the notion of an emotional residue left lingering. Remember, too, Claude Bowes-Lyon: 'If you could even guess the nature of this castle's secret ...'

10

RAYNHAM HALL AND THE BROWN LADY

'Captain Provand took one photograph . . . while I flashed the light. He was focusing again for another exposure; I was standing by his side just behind the camera with the flashlight pistol in my hand, looking directly up the staircase. All at once I detected an ethereal, veiled form coming slowly down the stairs. Rather excitedly, I called out sharply: "Quick, quick, there's something! Are you ready?" "Yes," the photographer replied, and removed the cap from the lens. I pressed the trigger of the flashlight pistol. After the flash and on closing the shutter, Captain Provand removed the focusing cloth from his head and turning to me said: "What's all the excitement about?"'

INDRE SHIRA, *COUNTRY LIFE*, DECEMBER 1936

'And the world will hail with delight the ushering in of that era when the interiors of men will be opened, and the spiritual communion will be established.'

ANDREW JACKSON DAVIS, *THE PRINCIPLES OF NATURE, HER DIVINE REVELATIONS AND A VOICE TO MANKIND* (1847)

HAUNTINGS

BY THE 1930S, THE OWNERS OF BRITAIN'S GRAND HOUSES WERE LOOKING AT THE glory days in the rear-view mirrors of the Bentleys they could barely afford to run. In truth, the rot had set in by the final quarter of the nineteenth century when foreign grain steadily undercut the price of that grown on the old estates. The First World War had taken unto God the sons of privilege as readily as it had the sons of toil, a generation of heirs and spares winnowed, like so much wheat, until the destinies of the grand houses from which they had sallied forth were blowing in the breeze, like chaff.

Many of the stately homes still standing and inhabited in the aftermath of that conflagration were laid low and lower by the Second World War. Requisitioned as hospitals, schools, and holding stations for soldiers bound for war in Europe and elsewhere, the passage through them of so much anxious and battered humanity in such a short time left their fabric gossamer thin at knees and elbows. They seemed done for.

It was in such an atmosphere of last gasp and desperation that threadbare lords and ladies had to look to their laurels. How were they to survive and retain ownership of their homes? Those with the stomach for the fight – or no other option – sought new ways to keep the lights on and the grass mown. It was looming hardship that summoned the dawn of the guided tour of the stately home, an uncomfortable trade-off that let in the riff raff (corralled by velvet ropes and kept moving by sprigs of holly on soft furnishings) but stopped the rot.

While the stately homes were in decline the religious movement of spiritualism – one of the most influential of its kind in the nineteenth century – was on the up. On 31 March 1848, in Hydesville, New York, fourteen-year-old Maggie Fox and her eleven-year-old sister, Kate, announced to a neighbour that they were in contact with a spirit they called Mr Splitfoot. They said he communicated with them by tapping from beneath the floorboards of their bedroom in response to any questions they asked. Messages from beyond the grave . . .

After some demonstrations in front of witnesses, the girls were sent away to live with their elder sister, Leah, in Rochester. Western upstate New York was a hotbed then of emergent faiths, and gave birth to Mormonism, Millerism and the earliest iteration of Seventh Day Adventism. So hot was the holy fire rolling back and forth over the territory, fanned by a cast of charismatic characters, like Jemima Wilkinson, the 'Public Universal Friend', and Andrew Jackson Davis, the 'Poughkeepsie Seer', that it earned the nickname the Burned Over District. Leah welcomed her sisters' newfound abilities and soon all three were on the road, filling theatres and salons with enraptured audiences. All too soon there were confessions of fraud – from Maggie, for a fee of fifteen hundred dollars from the *New York World*, followed a year later by a recanting she said her spirit guides had demanded from her. All three sisters died between 1890 and 1893, and despite controversies while they lived, the movement they had inspired survived them, and then some. A host of mediums had come forward, ready to intercede between the worlds of the living and the dead. A new religion was born and spread around the world.

It is worth noting that spiritualism was a draw during and after the internecine slaughter of the American Civil War, when unprecedented numbers of grieving families sought and often found comfort in a faith that promised to let them hear from lost loved ones. Having provided solace for the broken-hearted of one war, it would do the same for those similarly hurt by the butchery between 1914 and 1918. Shattered by loss, with many families grieving a father, a brother, a son, an uncle, the lure of connecting with the dead was irresistible for thousands. Faith in a God who would allow such a harrowing in the first place was rocked to its foundations. For many, the promise of someone – anyone – offering reassuring words from the Other Side took His place. With the better part of a million dead in just four years, Britain was a land haunted by loss.

It was as the twentieth century wore on that the owners of more and more stately homes were keen to attract paying visitors. Lord Montagu opened a motor museum at Beaulieu; the 6th Marquess of Bath installed lions at Longleat. And in a world altered by a First World War, and then a Second – in which a new faith pandered to those uninterested in established religion but curious about the possibility of life after death – other owners of grand old houses found it paid to draw attention to ghosts.

Raynham Hall, near the source of the River Wensum, near Fakenham in Norfolk, has in large part evaded the necessity to throw its doors indelicately wide. Home to the Marquesses Townshend for four centuries, it mostly remains just that: a private family home. The present incumbents, however, finance various restoration funds by holding music recitals and open days. One of the oldest stately homes in the county, Raynham is grand in scale yet still looks and feels like somewhere a reasonable person might imagine living. Building began in 1619 when Sir Roger Townshend, 1st Baronet, inherited his demesne at the age of seven. His father, John, had been killed taking part in the last horseback duel in England, on Hounslow Heath in London. Having had words with his kinsman Sir Matthew Browne, the pair rode out on to the heath on 1 August 1603 to settle the matter with their swords. Both were fatally wounded, Browne dying on the spot and Townshend the following day.

Sir Roger was a self-taught gentleman architect, inspired in part by Inigo Jones and his experience of a Grand Tour of the old classical world. He commissioned master mason William Edge to help him create a centrepiece for his estate and over the course of fifteen years they built perhaps the first home in England consciously to echo the Italian form and style. Students of architecture will spot elements drawn from elsewhere in Europe, including hipped roofs and Dutch gables. Surrounded by the green of informal lawns, the warm red brick of the hall's exterior is pleasing beneath a clear blue sky, primary colours working together.

The 800-acre park was laid out by Sir Roger's son Horatio Townshend, 1st Viscount. During his tenure, King Charles II visited Raynham in 1671, acknowledgement of Townshend's support of monarchy and restoration, and necessitating the styling of the King's Bedroom. There was fine-tuning in the 1730s when Charles Townshend, 2nd Viscount, employed Yorkshire-born polymath and designer William Kent for a refit. The fundamentals of the interior now are the product of his vision. The entrance hall, originally oak-panelled and dark, was made light and bright. Neo-classical Ionic columns have their echo in the Greek-influenced renderings of the four seasons, painted by Kent himself on the walls around the staircase. He was also responsible for the elaborately carved chimneypieces, decorated doorways and mosaic paintings. There is opulence throughout, but on a human scale that speaks somehow of homeliness.

Charles was a man of many parts. After years as an able politician in the House of Lords, he had grown tired of the cut and
thrust and withdrew to his country seat. There he became fascin-
ated by agriculture, and emerged as a vocal champion of crop
rotation and agricultural revolution. Such was his enthusiasm for
planting one root vegetable in particular that he was known lat-
terly as Turnip Townshend.

It was on 26 December 1936 that *Country Life* ran a story that
would establish Raynham as home to the world's most photo-
genic ghost. The piece had ostensibly been commissioned as
more of the usual content for a publication celebrating the grandeur of Britain's landed estates. Photographer Captain Hubert
C. Provand and his assistant Indre Shira had been despatched to
provide the accompanying images, and it was while they were set
up for a shot of the hall's grand oak staircase that Shira claimed he
saw 'an ethereal, veiled form coming slowly down the stairs'. He
called out to Provand, already beneath the focusing cloth, who
duly removed the lens cap to enable an exposure. And the rest
was paranormal history.

Provand allegedly bet Shira the image would show nothing
supernatural – but the photograph published in the magazine
appeared to show a ghostly feminine form floating down the
staircase. Instantly readers were writing in to declare the apparition must be that of the Brown Lady, by then long known as
Raynham's resident spirit. There were sceptics too, those who
said the 'ghost' was the effect of a smear of grease on the lens.
Others said it was a double exposure, the Brown Lady a photograph of some image of the Virgin Mary. To this day, the image
has its champions and its detractors.

Yesterday upon the stair . . . the Brown Lady of Raynham Hall

Those faithful to the legend say the Brown Lady is the ghost of Lady Dorothy – Dolly – Townshend, sometime mistress of the house. Born Dorothy Walpole, in Norfolk in 1686, she was a sister (reputedly the prettiest) of Sir Robert Walpole, Britain's first prime minister. By 1713 she was married to Charles, the 2nd Viscount, who later remodelled Raynham with the help of William Kent.

By 1713 Dolly and Charles already had history. Having met and fallen in love some years before, their intended union was thwarted by her father. Robert Walpole senior was also Townshend's guardian and feared people would say he was exploiting his position to secure a marriage from which he might benefit. Apparently, Charles was heartbroken and in haste married Elizabeth Pelham, daughter of 1st Baron Thomas Pelham of Laughton. The marriage produced five surviving children before Elizabeth died in 1711. Dolly and Charles were reunited soon after, and the pair picked up where they had left off.

Then, in a twist that, according to students of literature, helped inspire an entire genre of fiction, Charles learned that during their years apart Dolly had been the lover of the notoriously debauched nobleman and politician Thomas, Lord Wharton.

To make matters infinitely worse for Dolly, it was alleged by gossips that she continued to be intimate with Lord Wharton during her marriage to Charles – a union that nonetheless produced at least eleven children. Charles was known for his temper and, according to the legend, flew into a rage on hearing these allegations. Having dragged her to her rooms, he locked her in – and there she stayed, cut off from the outside world until her death in 1726. Just as there are those who disbelieve the photograph, so there are those who say that Dolly's death was faked too, her funeral a sham involving a coffin filled with bricks, and that her confinement lasted many years more. Perhaps she died of a broken neck in a fall, perhaps she succumbed to smallpox: there are all manner of rumours. No one seems certain. When Lord Raynham was interviewed by the BBC in 2009 he said: 'People said that Dorothy was locked away and badly treated, but in the 1960s we uncovered paperwork and medical reports suggesting she had a happy life and was much loved.'

Dolly's nephew Horace Walpole is credited with writing the first Gothic novel, in 1764, in the form of the masterpiece that is *The Castle of Otranto*. During the Victorian period there were more such offerings – *Jane Eyre*, *Frankenstein*, *The Woman in White*, often featuring a beautiful woman, wronged and locked away, violent, jealous husbands, women living in fear, old houses with hidden rooms, suspicious deaths . . .

Whatever the truth of the life and death of Dolly Walpole, the first recorded sighting of her shade came at Christmas 1835. There was a party and among the guests were a Colonel Loftus and a gentleman identified in the various accounts only by the name of Hawkins. At the evening's end, as the two made their way upstairs, heading for their bedrooms, they encountered an apparition, they said, a woman in a brown dress they described as old-fashioned, dated. Asked to draw what they had seen – gentlemen in those days were often schooled in the making of accurate sketches – one provided a work that prompted other guests to say they had had their own encounters with a similar figure and thought nothing of it, taking her for a servant. On a subsequent night, Colonel Loftus claimed to have a second sighting of the same, this time noting to his horror that the figure's eye sockets were but empty pits in a face strangely glowing.

In the aftermath of the men's testimony, more servants and others familiar with the hall came forward to admit that they had seen similar and had said nothing about it for fear of ridicule. Word spread and, for a while, the prospect of meeting the Brown Lady made Raynham a place to be avoided.

The following year, intrigued by the tale, a Captain Frederick Marryat arrived at Raynham determined to prove a theory of his own. A friend of Charles Dickens as well as of the Townshends, the incumbents at the time, he had a reputation as a pioneer of fiction set at sea. He was also a serving magistrate and had come to the hall primarily because he was convinced mischief was being made by smugglers operating nearby and keen to frighten away likely witnesses, the Townshends included.

What happened next was recorded by Marryat's daughter, Florence, and published in 1891 in her book titled *There is No Death*. Like her father, Florence was a prolific writer, mainly concentrating on novels best described as sensational, with titles like *Too Good for Him*, *At Heart a Rake* and *The Risen Dead*. With a keen interest in matters spiritual, she also attended seances organized by members of the spiritualist movement, and it was with her mind on the inhabitants of the other world that she offered a breathless account of her father's stay at Raynham.

He had chosen a room that featured a portrait of Dolly Townshend, depicted wearing that brown satin dress trimmed with yellow. Given his suspicion of smugglers in the vicinity, he kept by his side at all times a loaded revolver. The captain was undisturbed for two nights and then, on the third, two young nephews of Lord Townshend knocked at his door, asking if he would like to examine a gun they had purchased, newly arrived at the Hall from a maker in London. Dressed for bed as he was, he picked up his revolver and joked about it being a precaution in case of an encounter with who knew what? It was altogether a good-natured few minutes of company. The inspection of the new weapon over, the nephews offered to walk Marryat back to his room, joking about him needing their company in case of a ghostly meeting. It was as the three were nearing Marryat's room that they saw, ahead of them in the gloom, the approach of another. As Florence recorded:

> The corridor was long and dark, for the lights had been extinguished, but as they reached the middle of it, they saw the glimmer of a lamp coming towards them from the other end. 'One of the ladies going to visit the nurseries,' whispered the young Townshends to my father.

Since all had assumed they were about to meet a fellow guest and, since Marryat was in a state of semi-undress and therefore disinclined to greet a lady, all three ducked for cover into the nearest bedroom.

Decades later, Florence recalled her father's account of the next few moments:

> I have heard him describe how he watched her approaching, nearer and nearer, through the chink of the door, until, as she was close enough for him to distinguish the colours and style of her costume, he recognized the figure as the facsimile of the portrait of 'The Brown Lady'. He had his finger on the trigger of his

revolver, and was about to demand it to stop and give the reason for its presence there, when the figure halted of its own accord before the door behind which he stood, and holding the lighted lamp she carried to her features, grinned in a malicious and diabolical manner at him.

No shrinking violet, Captain Marryat was enraged and flung wide the door, took aim at the grinning face, and opened fire with his revolver.

The figure instantly disappeared – the figure at which for the space of several minutes *three* men had been looking together – and the bullet passed through the outer door of the room on the opposite side of the corridor and lodged in the panel of the inner one. My father never attempted again to interfere with 'The Brown Lady of Rainham'.

If there were subsequent sightings of the shade, they went unreported – until 1926 when the then Lady Townshend claimed her son and a friend had encountered the same vision, on the stairs once again and dressed all in brown. And there the story ended . . . for a decade.

It was the *Country Life* photograph that brought her back again, made a world-wide phenomenon of her indeed. Less well known is that the year before the image was captured, the sometime Lady of Raynham Hall, Gwladys, Dowager Marchioness Townshend, had brought out a book called *True Ghost Stories*.

Gwladys Ethel Gwendoline Eugénie Townshend (née Sutherst) was, in her time, a writer of novels, poems and plays, Lady Mayor of Kings Lynn and, by her own estimate at least, 'the first peeress to write for the cinema'. Criticized for coming up with movies that were 'overly melodramatic' – even by the standards of early cinema – she was undaunted and, between 1913 and 1915, produced eight silent films, including *The House of Mystery*, *A Strong Man's Love* and *Wreck and Ruin*.

The House of Mystery . . .

From 1905 until his death in 1921 she was married to John Townshend, the 6th Marquess. Their union was a scandal of the day, with numerous reports in newspapers that the impoverished peer had accepted Gwladys as his bride only on condition her father pay him £27,000. Sutherst submitted a deposit of £2,500, but only after the nuptials did it emerge that he was an undischarged bankrupt and unable to pay the rest of the money. Later, Gwladys could write about herself as 'The Girl Who Was Bartered for a Title by Her Father'. Scandal or no, the pair stayed the course, their marriage blessed by the arrival of a son and a daughter. There were even rumours that Lord Townshend was a lunatic – none of which were substantiated by the medical establishment of the day, who concluded only that he suffered from depression. Of her time at Raynham Hall, Gwladys wrote, in *True Ghost Stories*: 'I must confess I believe in ghosts and have for many years lived in a definitely haunted house.'

Gwladys subsequently married one Bernard Le Strange. He died in 1958 and she outlived him by a year. As for the Brown Lady of Raynham, she has barely been glimpsed since the famous photograph was taken. Those faithful to the legend say only that she has moved on, perhaps captured and carried away in the chamber of Colonel Provand's camera . . . Who could say?

More and more I struggle with the knowledge that my mum's mum died when she was hardly older than I am now. I was not yet born, so for me she has only existed as a figure in a handful of black-and-white photographs and in the few stories my mum has told me. In the framed photo my mum has near her bed she looks like an elderly lady. Her name was Margaret, but the family called her Peggy. I find it hard to feel a real connection to Peggy, my grandmother.

I knew her sisters, Jessie and Madge, my great-aunts. Both were alive into my twenties and loved me. Madge never married, was quiet, and quietly funny. When I think of her, she always has on her coat and hat, ready to go home to Glasgow. Jessie had more mischief about her. She married twice. Her first husband, Uncle Andy – also dead long before I was born – was an engineer. He was once contracted by Tate & Lyle to design a perpetual motion machine, another fact I struggle to accept, though I know that for years the blueprints of whatever he came up with were in our house, finally going missing in some or other flitting. Even while Madge and Jessie were alive, it seldom occurred to me that they were my missing grandmother's sisters. Peggy Neill, half of my mum and a quarter of me, is an unknown quantity and out of reach.

She outlived her husband, my grandfather. James Cameron Neill never made it out of his fifties either. So young, both of them, I realize now. James's mother died giving birth to him and, in the manner of the Victorian era into which he was born, he was christened over her coffin. His father remarried but James, it seems, did not get on with his stepmother. Unhappy at home, he jumped at the chance of volunteering for service in the First World War. Having lied about his age he joined the Marines and was in Gallipoli just long enough to be shot and badly wounded by friendly fire. Details are scant but it seems the bullet passed through one arm, through his torso, and then had to be removed from his other arm by the surgeon who patched him up. That was the end of James's war, and he was invalided out of the army and back home to Scotland while still in his teens. He was troubled by his wounds ever after and, despite his efforts to volunteer for the Second World War, declared medically unfit. His early death was attributed to those old hurts. I type those words and wonder how close I came to not existing. That bullet, a different path taken, no more James and no me. Never born at all is a lot less than a ghost.

After James died Peggy went looking for him at a spiritualist church in Glasgow. I asked my mum for details but there weren't many she could add. Her mum was lonely, she said, for her husband and also for her parents. Mum only remembered going with her once, to sit in the church and listen while a medium called out messages from the Other Side. There was never any word from James, but my mum recalled that the other churchgoers were kind, ordinary, decent people. One or two visited the family home from time to time and together they talked about anything but spirits.

11

CULLODEN MOOR, THE TWO SIGHTS AND DREAMS THAT WILL NOT DIE

'In August 1748, before the Town Council of Aberdeen, eleven men and women swore to the truth of a vision which they said they had seen in a valley five miles to the west of the city. On the fifth of that month, at two o'clock in the afternoon, they saw three globes of light in the sky above, which they first took to be weather-galls but which increased in brilliance until twelve tall men in clean and bright attire crossed the valley. Then were seen two armies. The first wore clothing of dark blue and displayed Saint Andrew's Cross on its ensigns. The other was uniformed in scarlet and was assembled beneath the Union Flag. Twice the red army attacked the blue, and twice it was beaten back. When it rallied and attacked for a third time it was routed and scattered by the Scots army. Those who watched saw the smoke of the cannon, the glitter of steel, and the colours waving, but they heard no sound. When the blue army was triumphant the vision passed.'

JOHN PREBBLE, *CULLODEN*

THIN PLACES WHERE THE SEPARATION BETWEEN OUR REALITY AND OTHER WORLDS is slight as gossamer, residues of emotions drifting like mist long after the people who first experienced them are gone, aching loss felt by millions all at once and summoning ghosts. The rational world is shot through, as Scottish scientist and psychical researcher Archie Roy says in his book *A Sense of Something Strange*, with 'things that over the years invoked within me a sense of something strange'.

Two years after the last pitched battle on British soil, the ugly horror of Culloden, witnesses swore they watched phantoms stage a rematch in a glen by Aberdeen. The Highlands of Scotland are worn thin by ancient hardship, patched with remnants of other days, other ways. Faith in the Second Sight – more properly the Two Sights – still lingers. Those so blessed (some would say cursed) see two worlds together, the here and now and times still to be. Often the visions come unbidden, day or night. Those most conspicuously gifted are called 'seers' and the most famous of all was one Kenneth Mackenzie, known also as Dark Kenneth on account of his sallow, swarthy complexion. Said to have been born sometime in the early seventeenth century at Uig, on the Isle of Lewis, he is mostly connected to the Brahan estate, seat of the earls of Seaforth, where he worked as a labourer.

He had his mother to thank for his visions. While taking a shortcut through a graveyard one moonless night, she met the ghost of a Danish princess. Before allowing the shade to pass by and return to her grave, Kenneth's mother asked it to grant her a wish: that her son be accorded 'the sights'. The ghost nodded her assent, and the following day young Kenneth found a peculiar smooth stone, an adder stone, blue and black, with a hole in the middle through which he could look and sometimes, sometimes, see the future.

The Brahan Seer he was then, and from the finding of that trinket until his terrible death he made all manner of predictions, many of which came true. As befits a figure so mythologized, there are those who doubt Dark Kenneth, or Coinneach Odhar as the Gaelic has it, ever existed. It is fair to say there are no records of his birth, but in the Outer Hebrides of the late seventeenth century, a place still in thrall to oral traditions, the absence of paperwork detailing another baby born into poverty is hardly surprising. As is the case with much of the past of the Highlands, a great deal was committed not to any page but to memory and only written down long after.

According to the myth spun about the Brahan Seer, he had foresight of the four lochs of the Great Glen made into one, a prediction fulfilled by the nineteenth-century construction of the Caledonian Canal. More than two hundred years before the railways, he saw black horses of iron, without bridles, breathing smoke and fire, hauling carriages through the glens. He saw 'black rain' make Aberdeen rich three hundred years before the discovery of North Sea oil and said Scotland would regain her parliament once men could walk dry-shod from England to France – a vision made real in 1999, five years after the opening of the Channel Tunnel.

Years before the Battle of Culloden he had cause to visit Drumossie Moor, which would be the scene of the fighting. For reasons known only to himself, the place moved him well before the fact. All at once he blurted out: 'Oh! Drumossie, thy bleak moor shall, ere many generations have passed away, be stained with the best blood of the Highlands. Glad I am that I will not see the day, for it will be a fearful period; heads will be lopped off by the score, and no mercy shall be shown or given on either side.'

If a country can be haunted by a battle, Scotland is haunted by Culloden. History is properly read by the few while the many make do – those who bother – with shorthand, bullet points and recycled propaganda. That clash on Drumossie Moor was, for the longest time, caricatured as a fight between Scots and English, between romantic Highlanders in their plaids and myrmidons of the House of Hanover in scarlet tunics. The truth was different, more complex, and recent generations of historians and archaeologists have battled to reveal Culloden for what it was: a brutal sideshow understood only in the context of another pan-European war. On an otherwise forgettable swatch of moorland close by the city of Inverness, part of the British Army, more preoccupied by the War of the Austrian Succession, enacted squalid butchery.

Provoked by the Hapsburg monarchy's failure to produce a male heir, most of the great powers of Europe were sucked into the maelstrom as Britain and her allies fought the tag team of France and Spain. Culloden was not a match pitting England against Scotland. Neither was it purely about Catholics fighting Protestants – Britain's most committed ally was Catholic Austria. With a view to stirring domestic trouble for the British government, and so distracting part of the British Army from business abroad, France took advantage of Scotland by blowing over old embers. It was a cynical political move that doomed thousands to ugly deaths.

Charles Edward Stuart is the figure most readily evoked by

Oh! Drumossie, thy bleak moor . . . *scene of the butcher's yard of Culloden*

mention of Culloden. He was Catholic, sent north with the backing of French Louis XV. Charles's grandfather had been James VII and II, a Catholic and a Stuart. He had been dethroned by his son-in-law and nephew, the Protestant William of Orange, in the so-called Glorious Bloodless Revolution of 1688. James's daughters by his first marriage, Mary and Anne, reigned after him, but when Anne died without an heir, in 1714, Parliament set aside the Stuart dynasty to make way for their German cousins of the House of Hanover. James VII and II had died in 1701, making his son, the would-be James VIII and III, the Old Pretender to the British throne. It was in his name that Charles, the Bonnie Prince, the Young Pretender, took leadership of the rebellion they called the '45 and sailed from France to the little island of Eriskay, in

Scotland's west, with seven followers remembered as the Seven Men of Noidart.

James II's supporters – calling themselves Jacobites, from 'Jacobus', the Latin form of the name – sought the return of the Stuarts and went to war on the matter four times. In 1745, with the War of the Austrian Succession in full flood, came the final roll of the dice. By then the war had spread into North America and the Caribbean. In the context of a conflict that had increasingly styled Britain as fighting the threat of Catholic tyranny, the Jacobites and their Stuart cause seemed inextricably linked with the wider interests of France and Spain.

There was support in theory for Charles among the clans – Catholic and Protestant, those likely to benefit from a Stuart return – but his arrival without French fighting men and ships meant his initial reception was cool. Influential voices advised him to go home – to which he replied that he was already at home. By force of personality and, crucially, with the support of Donald Cameron, the Gentle Lochiel, he drew enough men to his cause. There was a first luminous victory over the British Army at Prestonpans, and triumphant occupation of nearby Edinburgh, followed by invasion of England and advance as far as Derby, just 125 miles from London. There or thereabouts, Charles admitted he had had no word from English Jacobites – this despite claiming much earlier that they were rising and on their way. Confidence dwindled and, against Charles's express wishes, his force began the long walk home. There was an attempt to take Stirling Castle, but the garrison's guns were too much. With French soldiers in the mix at last, there was a final victory over a

British Army at Falkirk, but that slim success was followed only by more retreat all the way to Inverness.

On 15 April the bedraggled Jacobites marched through the night towards nearby Nairn in the hope of surprising their enemy camped there, in the aftermath of a party to celebrate the birthday of their leader, William Augustus, Duke of Cumberland, third and youngest son of George II and Captain General of the British Army. The attempt failed, ending in confusion and a withdrawal as far as Drumossie, with redcoats in pursuit. It was an exhausted and unfed Jacobite force that formed up in their clans to face the inevitable.

The rest is the stuff of grim legend: bombardment by British heavy guns that harvested them like grass. Pushed beyond endurance, those still standing set out across boggy, uneven terrain unsuited to their traditional downhill charge. The British heavy guns switched to grapeshot that made pink mist of hundreds. A clash of Jacobite targe and broadsword broke upon disciplined musket and bayonet drill . . . A bloody rout pursued by mounted dragoons.

There was talk of regrouping and fighting again, but it came to naught. The '45 was over. Charles had been escorted off the field and fled into the Highlands, then into the islands of the west. He was a fugitive for five months before returning to the mainland and sailing to exile in France, then Italy.

Those he had led and left behind were less fortunate. Of perhaps five thousand Jacobites who had stood to on the morning of 16 April 1746, some fifteen hundred were dead – either killed during the fighting or executed in the aftermath. Cumberland made much of Jacobite orders he said he had recovered on the field that stated no quarter would be given. True or not, the consequence of his reporting was the wholesale slaughter of wounded men on the field, then of prisoners taken to Inverness and elsewhere. The atrocity of the sixteenth was followed by a relentless and merciless harrowing of the Highlands – folk slain, homes put to the torch, livestock driven off. The Gaelic Highland way of life was uprooted and the ancient clan system all but destroyed, as was the intention. Britain was never again to be threatened by rebellion, thereby securing the foundation of an empire abroad.

Feted at first for his victory, Cumberland was known for a while as Sweet William. Soon enough, as the stories of murderous brutality circulated, the nickname changed to Butcher Cumberland. He never won another battle and, having once been touted as a commander of genius, subsequently resigned his post. No regiment of the British Army boasts Culloden among its battle honours. It had been dirty work, and those who had done it put it behind them as best they could.

As the Brahan Seer had foretold, so it had come to pass: 'bleak moor stained . . . with the best blood of the Highlands'. Many who visit the moor today attest to the residue there of so much hurt and cruelty. There have been innumerable reports of meetings with a tall man wrapped in plaid. He mutters to himself, under his breath, and those who draw close enough hear him whisper, 'Defeated . . . defeated.'

A famous account has a woman visitor to the field find a sheet of plaid upon one of several supposed mass graves on the edge of the battlefield, each marked with a rough boulder bearing the name of the clan in question. On lifting the cloth, the woman is confronted with the apparition of a dead Highlander that promptly vanishes. On the anniversary of the battle, the cries of men are heard, as well as gunfire and the sound of steel on steel. From time to time, armies of ghostly soldiers are seen charging and fighting on the moor.

A member of staff at the visitor centre recorded a visit by two ladies – one of whom lived nearby. They knew all about the ill-fated night march to Nairn by the Jacobites: the splitting of the force in the early hours of the morning as one of the two commanders opted to lead his men back to the moor, the perseverance of the second until an hour before dawn when news reached them of the order to return . . .

Those visitors had heard and read all of that but wanted the route of the march laid out for them. A map was produced, and when the line of advance and retreat was drawn, one pointed at a house on the map and cried, 'That's mine!'

On several occasions during her time in the house, she had been woken in the night by the sound of many men moving and marching close by. Every time she investigated . . . and every time there was nothing to see. Having been shown how the route of the march passed over ground that was now her garden, she was satisfied at last to know her nocturnal disturbances were the echoes across time of Jacobites making their way back to the moor and destruction.

Just as the Battle of Culloden left ghosts, it prompted more than the Brahan Seer to sense the coming horror. The Lord Advocate and Lord President of the Court of Session at the time of the '45 was one Duncan Forbes, whose ancestral home was Culloden House, overlooking Drumossie Moor. He was a staunch supporter of the Hanoverians and spent a fortune of his own money persuading clans not to rise for the Bonnie Prince but to remain at home. His efforts for the House of Hanover ruined him – financially, if not spiritually – for he was never properly reimbursed. Weeks before the climactic battle he was in Culloden House, looking out of a window and over the moor when – no seer he, as far as anyone knew – he said, 'All these things may fall out. But depend on it: all these disturbances will end on this spot.'

Gaelic folklore tells of the *bean nighe* – pronounced 'ben ni-yah' – the washerwoman. Deer that come down off the hills to walk the streets of villages are a portent of doom for Highlanders, forewarnings of death to come. So, too, the washerwoman, understood as the spirit of one who died in childbirth, like the mother at Vedbaek. Sometime before Culloden, Donald Cameron, the Gentle Lochiel, had a vision of the *bean nighe* washing burial shrouds in a stream. In this way, he knew, she was preparing for the dead of the battle to come. Lochiel would be badly wounded and carried off the field. Around him would fall scores of his Camerons, perhaps my ancestors among them.

Most unsettling of the portents, the harbingers of doom before Culloden, is a reported appearance by a terrifying otherworldly apparition called the Skree. Of all the experienced soldiers who supported and advised Charles, none was more able or well thought of than Lieutenant General Lord George Murray. The two men fell out before the end – most notably Murray urged Charles not to stand and fight on Drumossie Moor, declaring it hopeless terrain for Highlanders, who depended upon a downhill charge to create momentum and close the distance between them and their foe. Nonetheless, he was there throughout the '45, the mastermind behind the victory at Prestonpans and in charge of the skilful retreat from Derby.

On the evening before Culloden he was among the witnesses to an appearance in the sky above a company of his men of a great winged and hideous creature. It was shrieking, a noise fit to raise the dead. Described as like the harpy of Greek mythology, the apparition above Drumossie, beating black leathery wings, had a human head and blazing red eyes. Lord George Murray was no fanciful hysteric – rather, a sober and serious man.

Having foreshadowed the horror, the Skree has apparently neither forgotten the place nor left it in peace. In *Culloden Tales*, Hugh G. Allison recounts the experience of a battlefield tour guide identified only as Mike. He was with a party of tourists when his storytelling was interrupted: 'There was a full moon as we walked across the moor . . . Although the evening was cold, it was blessed by that winter's edge that makes you glad to be out . .'

He and his group were close by the area of the battlefield marked today as the position of the British Army front line when his attention was caught by 'what looked like a large black broken umbrella, straddling the path . . . Imagine our surprise . . . when it stretched itself and arose from the ground, looking like nothing quite so much as a giant black bat. It creaked its way into the sky, although not far, and then, after hanging there for a few seconds, it disappeared.'

He was adamant about the nature of its vanishing: 'It didn't fly away. It just disappeared, vanishing before our eyes, like the picture on an old TV . .'

He and his party searched the area for signs of the thing to no avail. The following day he reported what he had seen to the then manager of the visitor centre, who showed him a cutting from a newspaper, dated many years before, describing an encounter with a creature called the Skree of Culloden.

As for the Brahan Seer, his gift was his undoing. Isabella, wife of the Earl of Seaforth, summoned him and asked for news of her husband, then on a visit to Paris. At first he would say only that the master of the house was well. Isabella sensed, however, that he was keeping something from her and pressed him for details. If he would not obey her, she would have him killed, she said. And so Coinneach said what he could see, that the earl was with another woman, much fairer than she. Enraged and humiliated, Isabella condemned him to a terrible death – that he would be thrown head first into a barrel of boiling pitch. Before he died, the Brahan Seer said the Seaforth line would end with an heir who could neither speak nor hear. In due course, the seat was inherited by Francis Humberston Mackenzie, rendered deaf and speech-impaired by scarlet fever in his youth. He married Mary Proby in 1782 and the pair had nine children – five daughters and four sons. All of the boys died in childhood, along with three of the girls. As the seer had seen, the name died with those survivors, Mary Elizabeth Frederica and Helen Anne. It is said that while the old man had recovered the power of speech before his end, his last two years were spent in wordless mourning for his lost boys.

In an afterword of his account of the Battle of Culloden, in the daunted Highlands of Scotland, John Prebble wrote of how, in August 1748, two years after the gore, a party of men and women, eleven strong, stood before Aberdeen Town Council to swear the truth of what they had seen on the fifth day of the month. In a valley five miles west of the town they had watched, in the sky, three glowing suns that grew brighter until a dozen figures, straight and tall, crossed the ground in front of them in procession. Two armies formed up then – one blue and one red. Three times the red charged the blue, and three times was defied. There was the flash and smoke of gunfire, flags waving, but not a sound. After the last assault the red, under the banner of the Union, was routed altogether and the vision was no more.

12

The Tower of London, St Peter in Chains and Anne Boleyn

'Do not try to explain something until you are sure there is something to be explained.'

HYMAN'S CATEGORICAL IMPERATIVE

'In the Tower of London large as life
The ghost of Anne Boleyn walks, they declare
Poor Anne Boleyn was once King Henry's wife
Until he made the headsman bob her hair . . .'

STANLEY HOLLOWAY, 'WITH HER HEAD
TUCKED UNDERNEATH HER ARM', LYRICS BY
BERT LEE/ROBERT PATRICK WESTON

SOME STORIES HAUNT US LONG AFTER FIRST HEARING, CHARACTERS FIXED WITHIN and fossilized. Given no peace they are pored over, polished to a patina, or perhaps smudged and blurred, made indistinct and uncertain. Anne Boleyn is one who rests uneasy. Often maligned, her name is still trailed through mud: conniving temptress; husband-stealer; adulteress (incestuous adulteress at that); witch. Or else she is a figure to be pitied, a victim, which is worse. Worst of all is the damning with faint praise. Over and over, she is described as nothing much to write home about in the looks department (implication: why did Henry ever go there?), handsome at best, whatever that means for a young woman. A swarthy brunette in a world of gentlemen preferring pale blondes. Small breasts ('not much raised') when voluptuous was the order of the day. Nicholas Sander, an English Catholic priest, who was eight years old when Anne was beheaded so hardly a credible witness to her appearance, had nothing better to do than pick fault with her half a century later:

> . . . of sallow complexion, as if troubled with jaundice. She had a projecting tooth under her upper lip, and on her right hand six fingers. There was a large wen . . . under her chin, and therefore to hide its ugliness, she wore a high dress covering her throat.

That thing about the six fingers. Again and again, it is repeated – although no such observation was made about her while she lived. In his 1926 book, *A Short History of the Tower of London*, George Younghusband described an exhumation and examination of Anne's bones in the nineteenth century: '. . . there were the signs of a sixth finger on one hand'.

But given the febrile atmosphere at court, the constant gossip, the intense scrutiny, surely any deformity would have been noticed and discussed. At the time into which she was born – with all sorts of obsessions about witches and their giveaway physical signs, including the Devil's mark, extra digits, any sort of physical deformity – it seems hard to believe a king in want of a healthy male heir would have stooped to court and wed a woman with six fingers on one hand. Bad enough that she was from a middling family while he was a king and married to a Spanish princess.

Eustace Chapuys was imperial ambassador to England for Charles V, Holy Roman Emperor. Since his boss was also Catherine's nephew, it is hardly surprising Chapuys expressed loathing for her usurper, referring to Anne either as 'whore' or, little better, 'concubine', yet he never mentioned any deformity of her hand. As a staunch supporter of Catherine, surely he would have done so had it been the case. And when Henry was desperate to be rid, a deformed hand would have been evidence of witchcraft, a crime with which she was never charged. There has been talk too of a third nipple – another mark of devilishness – about which the king was similarly mute.

Sometimes it seems we might know less about her than we think. Despite the familiarity of her name, historians are not even certain of the date of her birth, so that in many ways the truth of her, the reality of her, remains elusive. Novelist Hilary Mantel wrote in her 2012 *Guardian* piece entitled 'Anne Boleyn: witch, bitch, temptress, feminist': 'Her image, her reputation, her life history is nebulous, a drifting cloud, a mist with certain points of colour and definition.'

Nebulous and drifting. A mist. How like a ghost already.

Thomas Cranmer and Thomas Cromwell undid Henry and Catherine's marriage. In January 1533 it was Anne's turn to warm the throne and Henry's bed, so a quick wedding took place before

Before the butcher's block . . . Anne Boleyn says her goodbyes

a handful of witnesses. She was six months pregnant with Elizabeth by the time she was crowned.

For her part, Catherine of Aragon is haunting too. Wife first to Henry's older brother, Arthur, she was widowed within months when he fell prey to the still mysterious sweating sickness. Seven years later Henry rescued her from obscurity and the dowdiness of dowager. He said he loved her, and maybe he did. When Henry was with his army in France in 1513, it was Catherine who rode north to address English troops bound for Flodden Field. When those men were victorious over the Scots, she sent her husband the blood-stained coat of James IV, who fell among thousands of his countrymen.

All those pregnancies, six at least, made her matronly soon enough, with only a daughter, Mary, to show for all her enervating toll. She was royal by birth, daughter of Isabella of Castile, and yet . . . and yet . . . in the end another woman cast aside by a man intent on his latest crush.

Henry's assumption of the role of head of the Church of England, and his abandonment and banishment of his lawful wife, horrified many. That he invoked Leviticus – claiming that since she had been his brother's wife, their marriage was unclean – was the cruellest cut of all. For the longest time, always perhaps, the great and the good preferred Rome and Catherine, yet he put her away out of sight. Her final home was Kimbolton Castle, in Cambridgeshire, where she donned a hair shirt and left her rooms only to attend Mass. She died, in January 1536, after sending a final missive:

My most dear lord, king and husband,

The hour of my death now drawing on, the tender love I owe you forces me, my case being such, to commend myself to you, and to put you in remembrance with a few words of the health and safeguard of your soul . . . I pardon you everything, and I wish to devoutly pray God that He will pardon you also. For the rest, I commend unto you our daughter Mary, beseeching you to be a good father unto her, as I have heretofore desired . . . Lastly, I make this vow, that mine eyes desire you above all things.

Katharine the Queen

In spite of Henry's demands, she refused ever to recognize Anne as queen. Though he had made her dowager once more – Dowager Princess of Wales – she called herself queen until the end and married until the end. During the embalming of her body her heart was found blackened in part, prompting rumours that she had been poisoned – perhaps on Henry's orders. More likely it was the damage wrought by cancer. On the day of her funeral, in Peterborough Cathedral, Anne miscarried a baby, a boy. Henry did not attend the service and saw to it that Mary, too, was forbidden to be there.

Henry was a man aware of the ticking clock. Two wives in and no sons. As with Catherine, he turned first cold towards Anne, then cruel. By marrying her man, Anne had, anyway, created a vacancy for a mistress. To quote another monster, Hannibal the Cannibal: 'And how do we begin to covet . . . by coveting what we see every day.'

Just as Anne had been one of Catherine's ladies-in-waiting, so Jane Seymour had performed the same role for Anne before catching Henry's eye.

According to commentators, Jane was nothing to write home about either. But in time she would do what no other could and provide him with a baby boy. Catherine had not gone quietly and neither did Anne, but Henry was in no mood for biding his time. Anne was convicted of plotting her husband's murder and on 19 May 1536, not half a year after Catherine's miserable demise, was decapitated. Her remains were bundled into a box left over from the storing of long bows, and the bloody mess buried in the Royal Chapel of St Peter ad Vincula ('in Chains') within the Tower.

We hear mention of the Tower of London and think of fear and imprisonment, torture and death. As much as anything, it is propaganda and myth that have so blackened the name of the near-thousand-year-old building. It was much else besides a prison: royal residence; part of the defence of London; a place of safety and safekeeping for the Royal Mint, public records and the Crown Jewels. But there is no doubting that during the sixteenth and seventeenth centuries it was also where enemies of the state were imprisoned, broken and executed. Having entered the place, many never left, not even in death – their bodies buried within the walls. Today it is a village of sorts, home to some of those who work in the Tower and their families. Down through all the years there have been stories of hauntings.

The Princes in the Tower, Edward V, aged twelve, and his nine-year-old brother, Richard, Duke of York, are said to have been murdered on the orders of their uncle Richard III. There have been countless reports of glimpses of the little boys, often in white nightshirts, holding hands or running in fear. Lady Jane Grey – queen for nine days after the death of Edward VI – was found guilty of treason and faced the block. On 12 February 1957, 403 years to the day after her execution, two guardsmen swore they saw her: 'a white shape forming itself on the battlements'. The ghost of her husband, Guildford Dudley, similarly despatched an hour before her, is to be seen weeping in the Beauchamp Tower. Margaret Pole, the sixty-eight-year-old Countess of Salisbury and truly the last of the Plantagenets, was condemned to execution without trial after two years of cruel imprisonment in the Tower. Refusing to go quietly, she kicked and struggled and had to be forced down over the block, where the first blow from the callow boy of an axeman gashed her shoulder. Breaking free she leaped up and ran, pursued by the axeman who struck her eleven times before she died. Her ghost is seen in a grisly re-enactment of her fate . . .

The list goes on and on.

One of the most famous hauntings occurred on a Saturday night in October 1817, when Edmund Lenthal Swifte, keeper of the Crown Jewels, was with his family in the sitting room of the Jewel House. In 1860, in answer to an enquiry in the Victorian journal *Notes and Queries,* he provided a full account:

> . . . the doors were all closed, heavy and dark cloth curtains were let down over the windows, and the only light in the room was that of two candles on the table. I sat at the foot of the table, my son on my right hand, his mother fronting the chimney-piece, and her sister on the opposite side. I had offered a glass of wine and water to my wife, when, on putting it to her lips, she paused, and exclaimed, 'Good God! what is that?' I looked up, and saw a cylindrical figure, like a glass tube, seemingly about the thickness of my arm, and hovering between the ceiling and the table: its contents appeared to be a dense fluid, white and pale azure, like to the gathering of a summer cloud, and incessantly rolling and mingling within the cylinder. This lasted about two minutes; when it began slowly to move before my sister-in-law; then, following the oblong shape of the table, before my son and myself; passing behind my wife, it paused for a moment over her right shoulder . . . Instantly she crouched down, and with both hands covering her shoulder, she shrieked out, 'Oh, Christ! it has seized me!' Even now, while writing, I feel the fresh horror of that moment. I caught up my chair, struck at the wainscot behind her, rushed up stairs to the other children's room, and told the terrified nurse what I had seen.

Swifte had no sooner swung his chair at the apparition than a sentry in the corridor outside was confronted by a figure he described as a giant bear with a human face. He thrust with his rifle, his bayonet lodging in a door.

As Swifte had it:

His fellow-sentry declared that the man was neither asleep nor drunk, he himself having seen him the moment before awake and sober. Of all this, I avouch nothing more than that I saw the poor man in the guard-house prostrated with terror, and that in two or three days the 'fatal result', be it of fact or of fancy, was – that he died.

Through it all at the Tower, behind it all, is Anne Boleyn. In 1864 one Captain J. D. Dundas reported watching a yeoman behaving strangely in the space associated with her beheading. The young man seemed ill at ease and turned, as though alerted by a sound or the sixth sense. It was then that Captain Dundas saw 'a whitish, female form' drifting silently over the stones. The yeoman barked a challenge at it, brandished his rifle with bayonet fixed, and promptly fainted. Fainting on duty was a serious matter and later the yeoman faced a court martial – but was acquitted largely on account of the testimony given in support of his claims by Captain Dundas.

Perhaps, though, it is the royal chapel in the north-west corner of the Inner Ward, by the Waterloo Barracks, that is the epicentre of sadness and of shame at the Tower of London. History records a place of worship there or thereabouts long before the coming of the Normans who raised the White Tower. The present structure was commissioned by Henry VIII but by the time of Victoria it was in a sorry state, neglected and moribund, so money and effort were expended to make good once more. For years its status as a royal chapel had been overlooked, and commoners interred within. To make room for their coffins, old burials had been disturbed and bones scattered. Restoration work began in 1876, removing rotten plaster to reveal older architectural details. The floor, uneven and subsiding in many spots, was lifted and relaid, any human remains boxed and reburied. Special care was taken in the chancel, the area around the altar, where the bodies of significant figures were buried after execution. Beneath a patch of paving measuring just six yards by four were thought to lie, among others, at least the following: Katherine Howard, who ran screaming Henry's name before she was dragged back to her imprisonment in Hampton Court prior to her beheading on Tower Green; the teenage queen of nine days, Lady Jane Grey; the author of her fate, her father-in-law, John Dudley, Duke of Northumberland; her father, Henry Grey, 1st Duke of Suffolk; St Thomas More; Robert Devereux, sometime fancy of Elizabeth I; and various Jacobite lords – William Boyd, 4th Earl of Kilmarnock, George MacKenzie, 3rd Earl of Cromartie, Arthur Elphinstone, 6th Lord Balmerino, and Simon Fraser, Lord Lovat.

James Scott, 1st Duke of Monmouth, eldest of Charles II's bastards and leader of the Pitchfork Rebellion of 1685 that cost him his head, was buried beneath the altar. Also in that dread space had been laid the remains of poor Anne Boleyn.

The stones above were lifted in October 1876. Efforts were made to identify the bones unearthed within but there is little certainty about the conclusions observers drew there and then. All were boxed and labelled with best guesses. Suffice to say it seems reasonable to imagine Anne's remains are among them.

If residues persist, of suffering and dying, then would not that chancel, where so many slaughtered souls lie bundled, be a likely spot in which to scent them? Perhaps each of us should pause there anyway and spare a thought for others who fell foul of the regime in their own times. We are surely most at risk when we allow ourselves to forget what, as a civilization, we have been capable of. The Tower of London is haunted by who we were.

After the disturbance and reburial of the remains, a captain of the guard was walking past the chapel at night when his attention was caught by light flickering in the windows. He pulled himself up on to a ledge, and when he looked within, he beheld a procession of the erstwhile great and good, knights in armour accompanied by their ladies. At the heart of the gathering, conspicuous in part by her slight stature, but mostly on account of her having no head, was Anne Boleyn. The captain watched, transfixed, as the figures filed along the nave and around the altar. After some minutes, darkness fell upon the interior once more and the ghosts, if ghosts they had been, dissolved into nothingness.

13

MARY KING'S CLOSE, THE WIZARD OF WEST BOW AND LITTLE ANNIE'S DOLL

'The beauty of Edinburgh ... consists chiefly in a quality that may be called abruptness, an unexpected alternation of heights and depths. It seems like a city built on precipices: a perilous city ... There are turns of the steep street that take the breath away, like a literal abyss. There are thoroughfares, full, busy and lined with shops, which yet give the emotions of an alpine stair. It is, in the only adequate word for it, a sudden city.'

G. K. CHESTERTON, *ILLUSTRATED LONDON NEWS*

ON 17 APRIL 2019, A GUIDE AT THE POPULAR EDINBURGH TOURIST ATTRACTION called Mary King's Close reported the disappearance of a little plastic doll from one of the rooms. According to the website of the company offering tours of the seventeenth-century street, buried beneath the city's Royal Mile, the toy in question was from the Daisy Doll airline collection designed by 1960s fashion icon Mary Quant. It had been purchased thirty years before in one of the

Beneath the City of Edinburgh – a mound of toys for a lost girl

nearby souvenir shops that attract tourists visiting the Scottish capital. What made it important enough to cause that guide to radio his boss and report its absence was that it had been bought by a psychic to comfort the ghost of a broken-hearted girl.

Mary King's Close is the most famous and talked-about part of Edinburgh's real, yet unreal, Underground City. For centuries, even millennia, the geology of Castle Rock was exploited by the people living in its shadow. When the Romans arrived in the latter part of the first century A D, they found the place dominated by an Iron Age tribe calling themselves Votadini, a name with Proto-Indo-European roots, meaning 'stand' or perhaps 'foundation'. The Votadini, later called the Gododdin, held a territory stretching from the Forth in the north to the Tyne in the south.

Castle Rock and nearby Arthur's Seat were obvious strong points for a people intent on domination. Arthur's Seat is the burned-out heart of a volcano, Castle Rock the plug of basalt that formed in one of its outlet valves hundreds of millions of years ago. When the glaciers of the last Ice Age spread southwards much more recently, even ice a mile thick was not enough to dislodge that doorstop of black rock. The unstoppable force broke around the immovable object and left untouched behind it a stretch of soft sandstone. The resultant geological feature is called, by geologists, a crag and tail. The crag would later be the foundation for the castle, the tail the steeply sloping Royal Mile. What the Votadini and others learned, and exploited, was that the tail was soft enough for burrowing, for the creation of shelters and caves. By the fifteenth-century iteration of the City of Edinburgh, that potential for subterranean expansion was well understood.

After the Scottish victory over the English at the Battle of Sark, in 1448, King James II gave orders to strengthen the city walls. The enclosure brought a sense of security against counter-attack, but also began a process of confinement. After the calamitous Scottish defeat at Flodden, in 1513, fear of full-scale English invasion occasioned the hurried raising of a new and greater wall yet. Now tens of thousands of people were confined in a space measuring a little over a mile long by just a quarter of a mile wide. When it came to building within, there were only two choices – upwards or downwards. Soon enough Edinburgh was home to the first skyscrapers, some fourteen storeys tall. Every square foot of available land was built upon, leaving just the meanest lanes and wynds threaded through the jumble, like constricted capillaries through fevered flesh. The vertiginous flanks of the tail were occupied too, by homes and businesses clinging to either side of steep lanes.

As a further defence against English attack, what had been pastureland at the base of the north face of the rock was flooded to create a feature known as the Nor Loch. Long since drained to create the famous Princes Street Gardens, in the sixteenth and seventeenth centuries the Nor Loch was the final destination for human waste and other foulness slumping downhill in open sewers from the buildings above. From its fetid shallows arose a Shakespearean pestilent congregation of vapours, a miasma that drifted into the wynds and closes, sometimes seeming to take vague forms suggestive of spirits from the other world. In the fourteenth century Froissart had called Edinburgh the Paris of Scotland. By the seventeenth it was a heaving cauldron of life and death and nowhere for the faint of heart.

Despite it all, or perhaps because of it, the city kindled and cradled greatness. English chemist John Amyatt is credited with having said of the place, in the middle of the eighteenth century, that it 'enjoyed a noble privilege not possessed by any other city in Europe'. He claimed: 'Here I stand at what is called the [Mercat] Cross of Edinburgh and can in a few minutes take fifty men of genius by the hand.'

For good or ill, the huddled masses shared that ripe stink of their own making. Those with money and titles occupied the middle levels of the taller buildings – above the stench but not so high as to demand too great a climb. Those without made do with whatever they could afford, at street level or even lower – in the spaces hollowed out of the rock itself. For those troglodytes, it was a half-life, an existence in darkness by night and overshadowed gloom by day.

HAUNTINGS

In *The Town Below the Ground*, Jan-Andrew Henderson described how the class system fought to survive in the midst of the mixing:

> A tradesman could, if he chose, build his house right next to titled gentry – and the poor squeezed in wherever they could find a corner. In the end, however, the class system prevailed . . . and the buildings themselves became divided up according to social strata. A single tenement in Dixon's Close was inhabited by: a fishmonger on the ground floor; a lodging-house keeper on the first floor; the Dowager Countess of Balcarres on the second floor; and Mrs Buchan of Kello on the third floor. The floors above these contained milliners and mantua-makers . . . and the garrets or attics were filled with tailors and every type of tradesman.

In the middle of the eighteenth century, work began on a fine new building on the north side of the Royal Mile – the Royal Exchange – in the forlorn hope of persuading merchants to conduct their business in more seemly circumstances than further down the hill and on the street. The plot chosen was directly opposite St Giles Cathedral, and the houses and businesses clinging to the slope below were repurposed as foundations for stately grandeur above. Those buildings that fronted on to the Royal Mile had their top floors removed and the lower transformed into the basement of the new structure. In the case of just one row of properties, however – Mary King's Close – the residents of the houses and businesses further down the slope (towards the malodorous morass of the Nor Loch) chose to remain in situ. Now they were truly and fully underground, topped off by the Royal Exchange squatting above them, like a cuckoo in their nest.

For years they stayed put, living and trading as before, but in a twilight world. During the second decade of the nineteenth century, the Royal Exchange was wholly abandoned by the mercantile class and converted instead into the City Chambers of Edinburgh Council. Necessary expansion later meant Mary King's Close had finally to be sealed up completely and, from then on, a street that had seen life of all sorts for hundreds of years was abandoned and entombed, out of sight and mostly out of mind.

In the absence of knowing, myths and stories took its place. I remember being told – at school, I'm sure – that during one of several outbreaks of pestilence in the city, infected people were shut inside their homes, doors and windows sealed against their cries. Legend had it that Mary King's Close was one of those places where this happened, and among the worst – stricken families condemned to terrible deaths, by starvation or plague, the howls of children muffled and ignored. In this way, we were told, the whole community was put out of sight, its inhabitants effectively buried alive.

Plague came to Edinburgh, and the sick were routinely confined to their homes, but no such horror as imagined for Mary King's Close ever happened. Nonetheless, the story stuck, adding another layer of fiction to the already fascinating facts of the city beneath the city.

Like Rome, Edinburgh is built on seven hills. Two are still proud – Castle Rock and Calton – but five are lost from view. Bunker's, Heriot, St Leonard's, Moultree's and St John's are still there, just engineered into invisibility. Edinburgh grew and spread, and to create more level ground, five high and mighty bridges were thrown across the valleys and ravines between the summits to make a tabletop of masonry. In this way the under-lying truth of Edinburgh was swept under a carpet of stone, a sheet thrown over the corpses of ancient dead. Now there was yet more buried underground. Above caves hollowed by the Vota-dini, above smothered streets, new voids were made by great arches holding aloft what they came to call the Athens of the North. How can anyone walk the streets of Edinburgh's Old Town and not wonder about the restless dead?

Mary King's Close was always accessible from the ground floor of the City Chambers, permission readily granted to organized groups. Since 2003 it has been an official tourist attraction, with a bespoke entrance on the Royal Mile. Stairs descend into a maze of fossilized passages, homes and businesses. Named after a fabric-maker who ran a successful business there after the death of her husband, Mary King's Close is a steep slope down towards the north (and the ghost of the Nor Loch), dimly lit for effect and with the clammy staleness of a coal mine. Among much else that is notable, if not haunting, is a glimpse of the lived reality of the poor. The finery of palaces and stately homes is all around for the gawping, but the soot-black, damp-stained, low-ceilinged walls of Everyman are less often preserved. The likes of Mary King's Close are where most of our predecessors came from, where most of us would have struggled to survive if fate had set us there and then instead of here and now.

Today, guides lead huddles of visitors through room after room, relating stories of known residents of every stripe, high and low. There are ghost stories, too, as might be expected in such a place. Numerous are the reports from visitors of walking into patches of intense cold, of feeling laid low by dizzying ague, of needing to return to the surface and the world of today. One of the last families to abandon the close were the Chesneys, who were saw-makers. Paranormal investigators claimed to encounter the presence of Andrew Chesney during a visit in the early 2000s and asked him if they should leave.

'Just get out,' he told them.

There's a lady in black; scratching behind walls attributed to a dead chimney sweep; the sounds of muffled chatter and clinking glasses – the whole gamut of manifestations that make Mary King's Close one of Scotland's most haunted locations.

Among the most infamous sightings is the ghost of one Major Thomas Weir, to be glimpsed, clad all in black, walking tall the length of the close to the place of his execution. In life, Weir had been a pillar of his community. He lived with his sister Jean, known to friends as Grizel, at an address in West Bow on the city's Grassmarket. The pair were respected as devout Calvinists. Weir was given to public prayer and preaching of the most impassioned sort, leaning all the while upon a black staff topped with a carved human head. Passers-by would often see him pull his hat low over his face whenever he had cause to pass Catholics or Anglicans, rather than look at them. A signatory of the National Covenant, he fought with distinction during the Civil War in the legendary army of the Marquis of Montrose. For all his undoubted idiosyncrasies, Weir was highly regarded, his reputation beyond question. When a woman accused him of immorality, while he was in his pomp, she was publicly whipped for her slander. And then . . . and then . . . in his early seventies and long retired, he abruptly confessed to a lifetime of Devil worship and witchcraft. He had had sex with mares and cows, he said, and he and Grizel had been lovers throughout their adult lives – a claim she loudly corroborated. She further declared that her brother had received his black staff as a gift from Satan, and that he had upon his flesh the mark of the Beast. She was marked too, she said, and proved it by frowning – whereupon an inverted horseshoe shape formed on her forehead. As a final flourish she said they had been transported together by a magical carriage, wreathed in flames, throughout the surrounding countryside.

Even by the standards of the seventeenth century, a place in thrall to talk of witches, the Weirs' claims were outlandish and dismissed at first as the ravings of madness. Andrew Ramsay, Lord Abbotshall, was provost of the city and insisted the major be assessed by doctors. Weir, however, persisted with his claims. Neither he nor his sister was ever tortured or otherwise coerced – their confessions being offered freely and unprompted. Both were convicted of incest and other wickedness. Grizel was executed near their home, but before they could hang her, she stripped off her clothes in a final defiant demonstration of her own declared depravity. Thomas was taken to a place of execution near what is now Leith Walk and there throttled with the garrotte, then burned. Before the sentence was carried out, he was invited to repent his sins. Weir shook his head and said: 'Let me alone. I will not. I have lived as a beast and I must die as a beast.'

His black staff was thrown upon the flames with him, and witnesses swore they watched it dance. As well as haunting Mary King's Close ever after, his spirit was said to linger in his old home at West Bow.

Weir's ghost might reasonably be expected to be the star turn for the subterranean tourists, but in fact that honour goes to the ghost they call Annie, the little girl in want of a doll. In 1992, back before the days of the formal tourist attraction, a Japanese psychic called Aiko Gibo paid a visit to the City Chambers while conducting research for a film about haunted places in Britain. Shown down into the spaces beneath, she was apparently unimpressed until she approached the threshold of a room in what had formerly been a seventeenth-century house. All at once she was struck by feelings of illness, she said, of unhappiness and misery. 'I cannot enter this room,' she said. 'It is too strong . . . there is a child beside me.'

Gibo reported feeling a hand tugging at her clothes. Finding the strength to step inside the space, she said she was in the presence of a broken-hearted child. She had lost her parents, and also her doll. Gibo at once returned to the surface and bought a replacement in a nearby shop. Since then, visitors with foreknowledge of Annie's plight have brought offerings of their own: dolls, teddy bears and other toys. These fill an alcove in the room and are periodically gathered up and given to local children's charities. It is never long before some visitor reports Annie's presence . . . an aching grief . . . loneliness . . . the touch of a little hand.

On 18 August 2013 my nephew Sonny was born. At the time of writing, he is nine years old. Soon after receiving the news of his birth, I phoned my mum and dad to share it with them. It was

Mum who answered, and when I told her the name my brother and sister-in-law had chosen for their boy, she was briefly quiet. I sensed there was something to know and asked her what. She told me again, in more detail, something she had mentioned more than once while I was growing up.

Mum's sister, Elma, was six years her senior. When Elma was three, my grandmother, Peggy, gave birth to a son. He was called James, after his dad – James Neill. Little James lived for just a few days. Sometime after the loss of him, Elma announced the presence of an invisible friend. The whole episode was over before my mum was old enough to be aware of it, and her subsequent version of events was based only on what Peggy told her later. The word Mum used to describe her mother's account of it all was 'vivid'. For Elma, her friend was real and very much there. He was in their lives every moment of every day for several years. Elma could describe him, hear his voice. A place had to be set for him at the dinner table, and a chair. When they travelled by bus or tram, space had to be allocated for him. He shared her bedroom. He was most definitely a 'he'. Elma said his name was Sonny.

Little James, the boy of a handful of days, was born on 18 August 1930, exactly eighty-three years before the birth of my nephew. Elma's Sonny was kindled in the aftermath of loss, appearing in the consciousness of a little girl too young to understand what had happened to her family. All this my mum repeated on the day of my nephew's birth. She made no more of it than the neatness of the coincidence – the date and the name – but it stuck with me. While writing this book I asked her about it again. She had no more detail to add, save to repeat that the presence of Elma's Sonny had been a vivid experience for all concerned.

NEIL OLIVER

Lost children are often in my thoughts, for reasons I struggle to understand. It's not a morbid preoccupation – more about awe. Over the years, the archaeological and TV projects I've been involved with have brought me into contact with ancient losses. Over and over, I have written about the little Viking they call Birka Girl, who died thirteen centuries ago on the Swedish island of Björkö. I think about three-year-old Suruchi Rattan, a little girl in red who was among those murdered by whoever blew Pan Am 103 out of the sky above the Scottish town of Lockerbie on the night of 21 December 1988. In the mid-1990s I was a cub reporter with the Dumfriesshire Newspapers Group. One of our titles, the *Annandale Herald*, was based in Lockerbie. My time on the staff coincided with the Lockerbie Inquiry and names like Suruchi's became all too familiar.

I have wondered if *Homo sapiens* is haunted by the loss of children, those events that are anathema to us as a species. I wonder if we are haunted by that which is most unthinkable, that worst of the worst that drifts from time to time into the peripheral vision of every parent, like a grey ghost. Do some of us carry the dread of it into places like Mary King's Close, and so find ourselves susceptible to encounters, imagined encounters at least, with a ghost like Annie?

As I get older I think more and more about Peggy, the grandmother I never knew. She lost her son after days. She lost her husband while she was still in her middle years and attended a spiritualist church in hope of finding him again. There are more than enough reasons for believing in ghosts.

14

BORLEY RECTORY, THE MOST HAUNTED HOUSE IN ENGLAND

'Yesterday, upon the stair,
I met a man who wasn't there!
He wasn't there again today.
Oh, how I wish he'd go away!'

WILLIAM HUGHES MEARNS, 'ANTIGONISH'

'Absence of evidence is not evidence of absence.'

ADAGE

BETWEEN 2013 AND 2018 I VISITED AUSTRALIA SO MANY TIMES I LOST COUNT. I was part of a team making the television documentary series *Coast Australia* for Rupert Murdoch's Foxtel, and the circumnavigation of that continent took some doing and multiple long-haul trips. For a while there, I had a platinum loyalty card from Emirates Airlines. As part of the filming of the fourth and final series we spent a week on the Cocos Islands, a few dots of

Australian territory adrift in the Indian Ocean, halfway to Sri Lanka. We told the story of the Clunies-Ross family, who were regarded as kings of the place from 1827 until 1978, and stayed for a single night in the home that had been the centre of their demesne, an underwhelming pile called Oceania House. According to the locals, Oceania House is haunted.

Originally discovered in 1609 by a British ship's captain, William Keeling, the Cocos remained splendidly untouched until the arrival of a Scottish merchant called John Clunies-Ross. His family hailed originally from the parish of Weisdale, on Shetland, but he was as far from home as it was possible for him to be when he stumbled on to his paradise. He was captain of the brig *Olivia* and sailing the Indian Ocean in 1825 when he happened upon the then uninhabited atoll of the Cocos, an archipelago of twenty-seven coral islands. Instantly enamoured – and quickly convinced there was a fortune to be made in growing and selling coconuts – by 1827 he had moved his family from Shetland to their new home. The Malay people he imported to be his workers were the ancestors of the six hundred or so Sunni Muslim Cocos Malays, who make up most of the islands' population today.

After that first John, four generations of his descendants maintained autocratic control of the Cocos Islands and the resident population. John's grandson, George Clunies-Ross, was master of the place when Queen Victoria granted the islands to the family in perpetuity. They remained in charge and in ownership until 1979 when, under pressure of expropriation, John Cecil Clunies-Ross sold the archipelago to the Commonwealth of Australia.

Oceania House was, for me, a place apart. Put simply, it did not belong: an incongruous imposition of the prosaic north on to and into a dreamlike southern idyll. Run most recently as a guest-house, it retains much of the feel of the time in which it was built, the latter part of the nineteenth century. The exterior is white brick imported from Scotland. Inside there is a great deal of dark wood. Home Island is one of only two permanently inhabited islands in the Cocos, and when we told any of the locals we planned to stay a night in the big house, they made plain (some in light-hearted tones, others less so) that they believed the house was home to spirits. It was unclear who or what was said to linger there after death.

In the early days of settlement, the conscripted Malay workers lived hard lives. HMS *Beagle* anchored in the Cocos for seven weeks in 1836 and Charles Darwin made notes about the inhabitants' condition. He felt they were 'discontented' and described how their settlement 'had a desolate air . . . They appeared poor, and their houses were destitute of furniture.' He wrote that at the time of his visit they existed 'nominally in a state of freedom . . . as respects their personal treatment; but in most other points they are considered as slaves'. Reading between the lines, it seemed Oceania House was seen, by the present-day inhabitants of the islands, to have drawn and held the spirits of the most put-upon of the early Malays. Their unhappiness had lingered. Those living on the island in the modern day, we were told, gave the place a wide berth. I wondered, too, if some or other of the Clunies-Ross clan, irked by their descendants' dispossession, had laid ghostly claim to the place.

Undaunted, we arrived and were greeted by our hosts – and left alone to do some filming for our documentary. Around the walls of the main hallway were mounted, like hunting trophies, bronzes of the heads of the patriarchs, each luminous bright with verdi-gris: the first John, looking like a Mr Rochester of sorts, his son, John George, with ZZ Top beard, then George and a Viva Zapata moustache, and John Sydney with lesser 'tache, as though any appetite for luxurious facial topiary had steadily diminished with the passing of years.

To walk past them was to witness a decline of sorts, as though each had had less power than his father and known it. The future had inched closer, and their outmoded ways made them ever more incongruous. Without a doubt, it is the oddest guesthouse I have visited. We were made welcome, for sure, but if any prepar-ations had been made ahead of our arrival, they passed me by. If Miss Havisham's mansion had been transported to the tropics and repurposed as a B & B, it might have been Oceania House. Everywhere was sullen with dust and the desiccated carapaces of insects long departed.

Suffice to say Oceania House is the only reputedly haunted house in which I have slept . . . or tried to sleep. My room was as dark as the rest of the interior – wood-panelled from floor to ceil-ing; my bed, as I recall, was a four-poster, although I admit that might be memory playing tricks. It was stiflingly hot, as might be expected on an island in the Indian Ocean in the summer. The night was oppressively flat, airless, as though under a pressure-cooker lid. Against all expectations, there were no mosquito nets – not over the windows and not around the bed. At first the only nocturnal visitations I was concerned about were those of the whining, ear-bound, airborne kind. I had had more than enough wine with dinner, and at first sleep came easily.

After some hours though (I am not sure how many), I awoke,

not into any kind of drowsy state that might have promised the prospect of more sleep, but wide-eyed, full consciousness. I felt at once that I was not alone in the room. Sharper yet, sharpest of all, I could not move. I wanted to reach for my phone and use it for a torch. It was on the floor beside my bed (there being no bedside table) but I was fixed in place beneath my sheet, slick with sweat. I might have expected to feel fear in moments such as those, but the truth is the sensation I recall most clearly is one of disbelief, or perhaps a sense that the whole thing was happening to someone else so that I was only an observer of my own predicament. After all, we had specifically been told the place was haunted. I had joked all day with my companions – a director, a cameraman, a sound recordist and a researcher – about likely ghosts. All of them, I knew, were in bedrooms nearby. I had even read, over the years, about the phenomenon of sleep paralysis, in which sufferers are awake but unable to move, often experiencing hallucinations at the same time. Even as it was happening to me therefore, the whole episode seemed too clichéd for words. And yet there I was, paralysed in my bed in a matt black room in a supposedly haunted house in which I could plainly detect the presence of another.

'Hello,' I said. It seemed ridiculous even to me, in that moment, and it crossed my mind that I might have been pranked by the rest of the team. I could see that the bedroom door was shut – in fact, there was a line of gold around it from a light left on in the hallway so no one seemed to have entered (surely, I would have heard as much). There was, anyway, no reply and I did not speak again. The sensation could only have lasted for moments and when I tried to move again, I was able to quite easily. I sat up, got down from the bed and reached for my phone. With its light I found the switch on the wall and turned on the overhead lamp in the centre of the ceiling. There was nothing and no one to be seen. I sat down on the bed. The feeling of another in the space was as gone as the dark. I lay down, phone in hand, but there was no more sleep that night.

As well as stately homes and royal palaces, it seems haunted reputations accrete also around unprepossessing architecture. When I first saw a photograph of Borley Rectory, in Essex – famously labelled by some the most haunted house in England – its essential drabness made me think of Oceania House. That neither building looks the part only adds to the impact when paranormal phenomena are alleged, like learning a maiden aunt has a taste for rap music or taking part in satanic rituals.

When it comes to Borley Rectory, photographs are all that remain. A house of thirty rooms atop a low hill overlooking the village of Long Melford and the market town of Sudbury, a mansion of sorts, built of red brick. Bland and gloomy, it had stables and a farm close by. Across the way, facing the rectory, was the church; Borley village all about, strewn carelessly across the Essex–Suffolk border. In 1939 a fire, judged by the insurers as having been started deliberately, had left the house a shell, and the upstanding remains were razed to the ground five years later. The site now is occupied by modern homes, built within the last sixty years or so.

At least one resident has claimed glimpses of dark shapes, electrical items switching off and on by themselves, objects moving of their own accord. If there's an atmosphere, it might be one of

Gone, but never forgotten – Borley Rectory

expectation. More than one visitor has claimed a sense of some-thing magical. From time to time there have been reports of mild seismic activity in the area. It is altogether odd.

The first priest at the infamous rectory was one Henry Bull, for whom the pile was built in the 1860s to replace an earlier iteration that burned to the ground. (It occurs to me that one rectory burn-ing to the ground on the site is bad luck, while two succumbing to the same fate, one after the other, might merit investigation.) The hauntings began at once, with sightings of a ghostly nun reported by some of the rector's daughters. An oft-recounted back story, that Borley was built over the ruins of a thirteenth-century monastery, may or may not have been generated by the girls. Home-grown or not, the legend had it that nearby there had been a convent, and that a monk from one and a nun from the other had fallen in love and planned to elope. Caught in the act they were put to death, she by being bricked inside a cavity in a wall of the convent. The ghostly nun was only the beginning.

When the elder Bull died, he was succeeded by his eldest son, Harry, who had grown up enjoying his sisters' stories. For a long time he was what used to be described as a confirmed bachelor, living a life of country pursuits and attending to the needs of his parishioners. Those sisters of his were much involved in the life of the parish as well and all was equable until later in life he married a younger woman – Catholic, to boot – with a daughter. The trio lived in a house across the road until his sisters abandoned the rectory for more seemly circumstances, whereupon Harry and his little family moved in and spent some quiet years, untroubled by ghosts of any sort. Finally, the upkeep became too much, and once the Bulls were gone, they were replaced by the Reverend Guy Eric Smith and his wife, a frail individual called Mabel.

It is hard to tell, from this distance, whether Mabel was weak in body, mind or both. Whatever the case, her state in that gloomy place was not helped by frequent visits from Edith Bull, one of the previous incumbent's sisters, who regaled her with stories of the hauntings. Early on, Mabel claimed to have found a human skull in a cupboard. Soon enough Guy Eric Smith contacted the *Daily Mirror* newspaper to report extraordinary happenings. The list of disturbances he noted became standard fare at Borley in the years ahead: more sightings of the nun (Bull senior had had a window bricked up to stop her looking in at him and his family having dinner), laboured footsteps coming from spaces known to be empty, unexplained lights in locked rooms, the ringing of the servants' bells when no one was near them, a phantom horse-drawn carriage given to vanishing in mid-flight across the lawns ...

Smith had, apparently, contacted the newspaper hoping to reach the Society for Psychical Research, but the journalists had had other ideas and sought instead the help of the pre-eminent ghost hunter of the day. Harry Price was a man of many parts, several likely homemade. Before making his name in the study of the paranormal, he worked (or at least claimed to have worked) as an archaeologist. He certainly sold greaseproof paper to butchers and contributed copy to more than one newspaper. At some point he had developed an interest in magic – conjuring, stage illusions and such. By the late 1920s his consuming passion, however, was the investigation of paranormal phenomena. Dismissed as a fraud in life and after his death in 1948, he was nonetheless associated with all manner of alleged hauntings. Borley Rectory was his most famous case by far, his masterwork if you will, and in 1940 he would publish his most infamous book: *The Most Haunted House in England: Ten Years of Borley Rectory.*

The Smiths left soon after their first reports of strangeness, but interest in the place persisted. The next tenants were the Reverend Lionel Algernon Foyster, a cousin of Harry, his much younger wife, Marianne, and their daughter, Adelaide. It was during their time in the house that the ghostly activity reached its frightening peak. Price returned to pick up the story and subsequently recorded more than two thousand incidents: spontaneous locking and unlocking of doors, objects appearing and disappearing, disembodied voices, a piano that played by itself, footsteps, glimpses of figures and other apparitions, more ringing of the servants' bells, strange odours, outbreaks of fire, objects of all sorts thrown around, tables and chairs overturned, injuries to persons . . .

There were any number of witnesses, but most interested parties acknowledged that Marianne was the focus of most of the activity. Then as now it was alleged, openly and discreetly, that for her own reasons – perhaps not least unhappiness with a much older husband, and living in a cold, draughty, hard-to-run house without electricity, gas or piped water – she was almost single-handedly responsible for the 'paranormal activity' at Borley Rectory. Perhaps the most famous of the phenomena was the appearance and disappearance of messages written on interior walls or on scraps of paper, almost always addressed to Marianne and requesting help or prayers. Later debunkers would say the walls were perpetually damp (on account of mortar made with sea sand, rather than the inland variety) so that words written on them would simply dissolve to nothing over time.

Lionel Foyster wrote three separate accounts of what happened to him and his family in the house. Perhaps, like Price, he had hoped to boost his otherwise modest income by producing a bestseller . . . Perhaps his young wife, unfulfilled in all manner of ways by her asymmetric marriage and by their dour, uncomfortable surroundings, kept herself amused by providing him with likely content. There was a lodger too, Frank Pearless, and years later Marianne would admit that she and he had been lovers during and after his time in the house. Perhaps some of the bumps in the night were altogether more earthly than spiritual, tall tales told rather than confess the carnal truth.

The Foysters left the rectory in 1935 after which almost all activity ceased. Either out of genuine belief or because he knew he was on to a good thing, Price continued to investigate, eventually employing people to stay in the house, in pairs, and record whatever unfolded. Some experienced otherwise inexplicable events but not in anything like the frequency of the Foyster years and never the same sort as before. Undeterred by the relative inactivity, and somehow in possession of Lionel Foyster's own writings, it was Price who produced the bestseller that propelled Borley into the status of national obsession. Eventually there were so many ghost hunters and sightseers prowling the place that the villagers took to removing road signs pointing in their direction. Anyone asking for the village from locations in the surrounding countryside was routinely sent on a wild-goose chase, deeper into the wilds, by those wanting only peace.

Borley attracts attention to this day. All manner of books have been written; countless online chatrooms host the faithful and the faithless. In 2000, a further attempt at a debunking was published as *We Faked the Ghosts of Borley Rectory*, by Louis Mayerling. Here the author himself is an enigma. While he claimed to have known the house from his earliest years – as a guest and as a friend of more than one family – at least some critics have doubted he was ever there and that he conjured the whole piece as self-promotion. Given his claim to have been, among other things, a lover of Marilyn Monroe, it seems fair to take his offering with a pinch of something as salty as the sand between Borley's bricks. Nonetheless, among the seemingly endless turns of the Borley story, Mayerling's account of having helped fake the ringing of bells, of making use of hidden passageways and secret doors, of seances, and of performing the role of the headless man sometimes glimpsed walking in the gardens, only add to the fascination.

That the building is no more is the final scintillating twist, putting the reality forever beyond reach. So many claims of strangeness by several families over many decades surely make it impossible, or at least unlikely, that everyone was lying all the time. The experiences of the younger Bulls, the Smiths and the Foysters unfolded, in whole or in part, between the wars, the hardest of years. A whole population shell-shocked by the horror and loss of the First World War, the great and the good brought low, as well as the middling and lower classes, reduced circumstances born of economic hardship and then financial collapse. It was also the time of that resurgence of interest in spiritualism, when uncounted numbers opened their hearts and minds to the possibility of contact with the dead.

At Borley Rectory, one family after another found it hard to get by in a dilapidated, sprawling house lacking even basic amenities. In earlier years the readily affordable presence of servants and gardeners might have made it bearable, even pleasant; in the straitened circumstances of the diminished inter-war years, it was a tough furrow to plough. Hard existences in the shadow of loss and grief might give rise to anxieties and doubts that manifest in all manner of ways in the minds of those with cause to fear life itself.

I had my own experience in Oceania House, on the Cocos Islands, at the other end of the world. As for Borley Rectory, gone without trace, I have only the words of others. Among all his enthusiastic debunking, Louis Mayerling claimed to have been present at one incident that left him shaken. During the Easter of 1935 he took part in a seance at Borley attended by T. E. Lawrence, Sir Montagu Norman, governor of the Bank of England, George Bernard Shaw, Bernard Spilsbury, Home Office criminal forensic scientist, and Marianne Foyster.

'We chose an ill-lit and underground cellar at about midnight and sat in silence. Someone gave a nervous cough and was about to speak when an extraordinary thing happened. The kitchen bells seemed to clang together in one single clash.' According to Mayerling, there were no others in the house at the time.

'Norman jumped up and then there was a lightning strike of silver-blue light, which appeared to implode from all the walls and the ceiling of the cellar and then there was a dead silence.' He recalled how everyone present was frozen to the spot, as though struck by a momentary paralysis. He claimed to have been blinded, only recovering sight in one eye. 'I can't explain that occurrence,' he said. 'And, to be honest, it still makes me feel rather shaken. The rest of the hauntings were, without exception, the most successful hoax of the age, but that still sets my spine tingling.'

15

WISTMAN'S WOOD, DEVON, AND THE DEEPEST ROOTS

'The past is never dead. It's not even past.'

WILLIAM FAULKNER, *REQUIEM FOR A NUN*

'If we had not met, that day, I think I would have imagined you somehow.'

PENELOPE LIVELY, *CONSEQUENCES*

WE ARE HUNTED AND HAUNTED BY DEATH. I ARRIVED AT MY PARENTS' HOME IN Dumfries a couple of hours after my dad died. He had been in the process of dying for some few weeks by then, so we were only waiting. He had passed the point of no return, as they say, was silent in his bed and drifting. I was at my home in Stirling when my sister called. No more than half an hour earlier a thought had occurred to me, and I had walked from my study into another room to find my wife and share it with her. 'I won't ever talk to my dad again,' I said. 'I've had all his words.'

'That's so sad, Neil,' said Trudi.

I walked into the kitchen to make a cup of tea – mostly for something to do. Before the kettle boiled, my mobile rang. 'He's gone,' said my sister. 'Dad's gone.'

Life is never quite the same after you have looked into a parent's face when life has gone from them. I see my own differently in the mirror now. I had been seeing my father's face ghosting behind my own for years, rising to the surface, but this is different. *Timor mortis conturbat me* – the fear of death torments me. Now I know what that face looks like after the light has gone out.

The Wild Hunt is a haunting, haunted tale that winds and trails across the folklore of Britain and Europe, drumming hoofbeats and the howl of hounds. From west to east it draws upon, in one location after another, a cast of characters meaningful and memorable to each place in turn. Like a folkloric format, it provides a narrative arc into which may be inserted the persons and events deemed to have mattered to the place and the folk in question. Local heroes and villains are twisted through all the countless tellings and retellings so that countless places, usually remote, high, lonesome locales on the fringes of civilization, are claimed as terrain over which ghostly hunting parties ride, mounted on bloody-hoofed horses, black-horned he-goats, and led by red-eyed dogs.

The Wild Hunt is surely in the Nazgûl, the Dark Riders in

J. R. R. Tolkein's *The Lord of the Rings*, being nine wraiths of what were once nine men, kings of Middle-earth, but diminished and enthralled by the power of the One Ring. Endlessly and tirelessly, they ride in its pursuit. In Wales the pre-eminent huntsman is the hero-trickster Gwydion, or else Gwynn ap Nudd, the king of the fairy folk, who drives ahead of him the Cwn Annwn, the Hounds of Hell. In Ireland there is Nechtan, king of the gods that reigned there before the coming of Christianity. In England the leader of the pack is variously King Arthur, Herne the Hunter, Odin, Herla, legendary king of the Britons, or the Devil himself. There are more names besides.

Wherever and whenever it originates, the tale is certainly ancient, and unifying to us as a northern, western variant of our species. There is no single form, no definitive recipe, but among the ingredients likely to feature are memories of battles, reverence of the ancient dead, remembrance of what was lost when hunters turned farmers, kings and queens as hunters and hunted, and, above all, faith in a slumbering hero who will rise and return when his people are most desperately in need. There is, too, a recurrent connection to those routes, sometimes called coffin roads, over which the dead were once carried towards their final resting places. On such tracks and trails, the traveller and wayfarer may expect to encounter the spirits of those dead, endlessly repeating their last journeys.

Always and everywhere, a glimpse of the Wild Hunt is a presentiment of doom. The ghostly riders have as their kin the Four Horsemen of the Apocalypse – famine, pestilence, war and death – and at the very least the soul catching sight of the Hunt is said to have been forewarned of his own imminent demise.

The story is everywhere in the mythic past of Britain and northern Europe, haunting all. It is never far beneath the surface either. Somewhere in that home of my parents there is a copy of *The Wild Hunt of Hagworthy*, by Penelope Lively, about the terrifying consequences of a village vicar's plan to revive the tradition of the Horn Dance. It was one of a slew of books I read in my early teens – *The Owl Service* by Alan Garner, *Stag Boy* by William Rayner, *Billy Buck* by Josephine Poole, *The Dark is Rising* by Susan Cooper, a long list of titles straddling the boundary between adolescence and adulthood. *Stag Boy*, *Billy Buck* and *The Wild Hunt of Hagworthy* featured deer, one way or another, and drew upon the same kind of folklore that enshrines the Wild Hunt.

The contours of the south-west-facing slope of the valley that cradles the West Dart River, in Dartmoor, Devon, are softened by a scrap of ancient oak wood. This is Wistman's Wood, tattered remnants of the forest that covered the south-west of England (and much of the rest of the British Isles) for millennia before the advent of hunters armed with stone axes ten thousand years ago. Surrounded by sheep-cropped moorland, Wistman's amounts to just eight acres – four football pitches' worth of tree cover. To be precise, it is a survival of temperate rainforest, a habitat that is rare not just here in Britain now but around the world. It is a fairytale landscape of so-called dwarf oaks, twisted and stunted like escapees of the bonsai keeper's art, the vicissitudes of the upland climate discouraging any lofty aspirations. Instead, the tallest trees of Wistman's – there are hawthorns, hazels, hollies and willows too – are little more than twice the height of a man. Rather than reach skywards, the trunks and branches hug the ground as though suffering from vertigo, lying there and clinging on for dear life.

Wistful ways in Wistman's Wood

Mosses and lichens need no asking twice, enfolding the trees, and everything else, draping them in dripping dampness. All manner of dependants – epiphytes all, vascular plants of winsome tenderness – thrive in greater profusion than might be expected in woodlands closer to sea level, polypody ferns, liverworts and more. Green is everything – and in a thousand, thousand shades: green the strewn boulders, green the fallen branches in decay. Earth and decay, sweet decay, are the scents that fill the settled, somnolent air.

Magic is as elusive as any ghost, but for those in search of it, it is surely there in Wistman's Wood, a name that likely has its twisted roots in *wisht*, meaning 'eerie' or 'uncanny'. Another word scarce-used now is *seelie*, derived from the Old English for 'happy', 'peaceful'. Its opposite is therefore *unseelie*, descriptive of something, or someone, wretched and perhaps wicked as well. Where both survive now is in connection to the fairy realm – split into a seelie, or good, court and an evil, unseelie, one bent on mischief. The folklore of Dartmoor has Wistman's Wood as home to the unseelie court, where a Wild Hunt from the Otherworld bides its time between rides. The wild rumpus there is often led by a dark giant, cloaked in black with a spear for his weapon and a horn for controlling his hounds and chilling the blood of the hunted. Infants are the quarry, taken away and replaced with changelings, in the manner of William Allingham's 'The Fairies':

> They stole little Bridget
> For seven years long;
> When she came down again
> Her friends were all gone.
> They took her lightly back,
> Between the night and morrow,
> They thought that she was fast asleep,
> But she was dead with sorrow.
> They have kept her ever since
> Deep within the lake,
> On a bed of flag-leaves,
> Watching till she wake.

On Dartmoor a story tells of a man making his way home from Widecombe Fair, through Wistman's after dark. From out of the shadows spring the hellhounds of the Wild Hunt, driven hard by the dark man. In hopes of keeping the peace and saving his skin, he shouts, 'How is the hunt?'

Without slowing, the huntsman turns to face him, and replies, 'Take that!' and throws him a leather sack. The man catches it and runs for home, arriving safe enough. Reunited with his wife, he opens the sack and finds inside the dead body of their infant son.

Sir Francis Drake – his name has its roots in *draco*, and so 'dragon' – was rumoured to have been in league with the Devil. Such was his success, the daring of his exploits, it was assumed by some that his gifts were otherworldly. His Devonshire home was Buckland Abbey, and for reasons no one now remembers, a version of the Wild Hunt of Wistman's has Drake as the leader of the hounds. According to the local legend he targets the unbaptized and the unwary, and in any event, there are those who claim to have seen him and his pack issuing forth from among the trees.

All manner of dreads are there in the Wild Hunt, as though it functions as a reservoir for dark thoughts and deepest fears.

The association with the coffin roads is a constant, with the huntsman often said to be looking for or fleeing from his own grave. The Lych Way is a medieval coffin road leading from the West and East Dart valleys to what was for a long time the only consecrated ground for Christian dead, at Lydford. Before Widecombe parish church was so blessed, at the end of the thirteenth century, the way was traipsed by burial parties. As it happens, the Lych Way passes just to the north of Wistman's Wood and even after all the centuries since it was last shadowed by a passing coffin, some claim to have glimpsed ghostly processions. Eventually the two ideas merged, and now the Wild Hunt is often about the restless dead, jealous of the lives of the living.

Dartmoor was also the setting for Arthur Conan Doyle's *The Hound of the Baskervilles*, inspired, according to some, by stories woven around the life and death of Squire Richard Cabell of Buckfastleigh. Apparently, he lived for hunting and was also described, even while he lived, as a 'monstrously evil man' who had sold his soul to the Devil. He died on 5 July 1677 and was laid to (un)rest in the family mausoleum at Holy Trinity Church in Buckfastleigh. That night a pack of spectral hounds came across the moor and circled his tomb, baying and howling. Ever after, so the story goes, on the anniversary of his death, he leads the pack once more. The locals went to some lengths to end his roamings and Sabine Baring-Gould's *Little Guide on Devonshire*, published in 1907, has it that

> Before the S. porch . . . is the enclosed tomb of Richard Cabell of Brooke, who died in 1677. He was the last male of his race and died with such an evil reputation that he was placed under a heavy stone, and a sort of penthouse was built over that with iron gratings to it to prevent his coming up and haunting the neighbourhood.

Wistman's is often described as only in decay and decline, but in recent years it has regained old ground, regenerating in spite of the sheep, spreading its roots and springing new life. If ghosts are mostly hidden, out of sight of the living, those trees, that place, might be a ghost of Britain past, lurking and sulking in the protective embrace of an unexpected river valley. Always the river makes its presence felt. 'Dart' is from Old Devonian for 'oak', and it is with the oak tree that the Druids, hooded and haunted, are most closely associated. 'Druid' might be from two Proto-Indo-European words – *deru* meaning 'oak', and *weid* meaning 'to see or know'. Dart, Druid, oak, all rising from the past and touching the present, like ghosts, imbuing Wistman's with ancient magic and fear.

There are few stories anywhere with deeper roots than that of the King of the Wood, and it haunts Wistman's Wood as well. Scottish anthropologist James George Frazer made it the tap root of his seminal work, *The Golden Bough*, and described the ancient rite by which a priestly king (or kingly priest) reigned alone within the confines of a sacred grove of oak trees. Never did any king's head rest as uneasily, for he was subject always to the threat of being slain by a would-be usurper, and slept fitfully, sword in hand. Any runaway slave might seek the sanctuary of the grove: by seizing an oak bough – a golden bough – he was entitled then to challenge the incumbent and, by killing him, replace him as king. Perhaps every king had been a slave first. Frazer interpreted the legend as the product of an ancient fertility rite for farmers, whereby a 'king' was sacrificed each year to ensure the fertility of the land. There it is again – that fixture of the Wild Hunt, the notion of a hunted king, forever restless and filled with the need for vengeance. Those farmers remembered always their hunter ancestors and were ready to sacrifice one of their own, a king, upon the altar of Diana, goddess and huntress.

By Wistman's flows the Dart, life in the midst of decay. The river is a symbol of renewal. Heraclitus observed the impossibility of stepping twice into the same river and in 'Sonnets from the River Duddon: After Thought', William Wordsworth wrote:

I see what was, and is, and will abide;
Still glides the Stream, and shall for ever glide;

The Form remains, the Function never dies . . .
And if, as toward the silent tomb we go,
Through love, through hope, and faith's transcendent dower,
We feel that we are greater than we know.

In days troubled by pandemic pestilence, war and death, Wistman's Wood and the Wild Hunt are reminders that the Riders of the Apocalypse are always in our peripheral vision. As a species it seems we are never without them, can neither escape them nor let them go. At Abbot's Bromley, in Staffordshire, there is a horn dance on the first Monday of September. Against the backdrop of modern houses and shops, dancers carrying reindeer-antler headdresses dance their way all day. It is a happy event that draws tourists, but something older lingers, something pagan and never laid to rest. The antlers have been radiocarbon-dated to the eleventh century and no one now remembers where they came from, or why.

Wistman's Wood is a shade of our own past, a reliquary holding safe all manner of needs and wants: the guilt felt by farmers for a way of life abandoned; the fate of old kings, doomed always to be replaced by youth; pagan rites and roots deeply buried.

Like ignorance and want, ghosts have been with us since the beginning. The English word has roots in Old English *gast* and also German *geist*, which suggests 'spirit' and also 'mind'. *Gast*

referred specifically to the breath, wherein the essence of life was suspended. A person's very soul was believed to exit the body in the final breath, the exhalation drifting unseen then, among the living, even entering and possessing the body of another.

From the beginning of written history, in the Old World of Mesopotamia, Sumer, Babylonia, Egypt, Akkadia and Assyria, there were variations on the theme of ghosts existing within an underworld community structured not unlike that of the living. Poor people expected that in death they would continue to exist at the bottom of that ghostly society. The rich commanded their slaves' deaths should follow swiftly upon their own, that those souls might serve them throughout eternity.

Pharaoh Amenhotep IV of Egypt sought to replace the traditional pantheon of Osiris, Horus, Set and the rest with a single god, One God, called Aten. He renamed himself Akhenaten, meaning 'useful to Aten', and tried to persuade the population he was that god's living embodiment. Despite his efforts, his monotheism did not outlive him. The people preferred their old faith and their old gods, and once Akenhaten was dead the priests of the old religion placed a curse upon his name, condemning him to an eternity as a restless ghost, haunting the desert.

The Greeks of the classical world imagined faint shadows of the living, insubstantial, weak and all but powerless. Sometimes those shades might appear bearing the wounds that had caused their deaths. Ghosts haunted the Romans too. Pliny the Younger wrote about a house where the clanking of chains disturbed its residents until investigation revealed a shackled skeleton buried beneath the floor. Once the chains had been removed and the bones reburied, there was peace.

All around the world there are descriptions of disembodied spirits and souls lost on the way to the next world, or to Heaven, or remaining behind to seek vengeance for wrongs suffered. Among the Japanese there are *yurei* – literally 'faint souls'. The *yurei* must make their way to the realm of the ancestors, where they join forces to protect those family members still alive. When a violent death has been inflicted upon them, the *yurei* must wait until those left behind have performed the appropriate rituals to help them move on. Something similar exists for the people of India, who describe *bhuti*: like *yurei*, they depend upon the performance of burial rites to secure their passage to the next world. In the Chinese tradition there are *gui*, and also *gweilo* – a word that means 'ghost' and 'foreigner', both pale.

While filming a documentary in New Zealand's North Island I arranged to meet a Maori elder at a headland on Poverty Bay called Young Nick's Head. The Nick in question was a twelve-year-old boy aboard Captain James Cook's *Endeavour* who, having been first to spot the landmark, had the honour of knowing it would bear his name ever after. Long before those Europeans arrived in the North Island of New Zealand, the headland was Te Kuri a Pawa – Pawa's dog – named in memory of a dog lost nearby by a hunter and believed to be there still, awaiting his master's return.

I was travelling with the film's director, Chas Twogood, and we arrived ahead of the rest of the crew. I left Chas in the car making calls on his mobile phone and wandered down to the shoreline. It was a perfect calm day – blue sky, warm, not a breath of wind. I walked close by the water. There were no waves at all, the surface flat and still as oil. I had the unmistakable feeling that someone else was close by. Assuming it was Chas, I turned to speak to him. No one was there, or anywhere in sight. I looked towards the parked car, some hundreds of yards away, and saw Chas where I had left him. The feeling of another's presence was still strong. The hairs rose on my arms and neck, and my skin prickled as though with cold in spite of the sun's warmth. I heard a vehicle engine and turned to see the crew arriving in their car. The sensation stopped.

The Maori elder arrived soon after, and on camera, I talked to him about the significance to his people of Te Kuri a Pawa. He explained it was one of a few locations around North Island where tradition had it that the souls of the dead would gather for a while, as though in an airport departure lounge, readying themselves for the journey back to the mythical home island of Hawaiki where Io, the supreme being in Maori religion, had summoned the world into being long ago. Traditional Maori believe that every person comes from Hawaiki, and so returns there after death.

We listened to the elder tell the story, and after we had finished, and while the crew put the kit back in the car, I told him about my experience on the beach. 'I had the distinct feeling that someone else was there,' I said.

The elder looked at me, green eyes shining.

'Oh, people were there, all right,' he said. 'The people are always there.'

For the second time that day, my hackles rose, and they are rising again as I type these lines.

16

WAYLAND'S SMITHY

'All that you've loved is all you own.'

TOM WAITS, LYRICS BY KATHLEEN BRENNAN /
THOMAS ALAN WAITS, 'TAKE IT WITH ME'

FOR THE LONGEST TIME, GUILT AND REGRET HAVE BEEN PART OF WHAT IT IS TO BE human and alive. When our species, *Homo sapiens*, arrived in Europe and the Middle East they displaced Neanderthals, their elder relatives. Cuckoos in old nests: Esau and Jacob all over again. Pushed from the hunting lands and on to the margins, the last of the older folk might have inspired the northern European stories of trolls – half-glimpsed lumpen figures, ever fewer in number and keeping to the shadows and dark places out of sight. Then, in their turn, *Homo sapiens* hunters were displaced by land-hungry farmers, their own kin, and driven towards the same diminishing and oblivion they had inflicted upon their elders. Over and over, something taken, and something lost. Here today and gone tomorrow.

Some of our kind understood early on that their existence exacted a price from the world: taking animals for food and hide; clearing woodland and burning trees. The miners who dug great shafts into the Earth at Grime's Graves, in Norfolk (an Anglo-Saxon name that means 'the diggings of the hooded man', who may or may not be Woden, or Odin), left food, and figurines carved from the chalk, as votive offerings, in payment for the flint they had taken from the world below. Debt paid, or at least acknowledged, they back-filled their workings and walked away, wondering if they had done enough. In the ages of metal – bronze and then iron – swords and torcs, items of inestimable value, were bent and broken and given back to the Earth or the water, back to gods or God. Somewhere an account was kept of what was owed, what was to be repaid. We are each borrowings from the Earth and from our parents' flesh and blood. Everything we are, held only for a little while before the account comes due.

The French anthropologist Claude Lévi-Strauss said mankind knows itself to be born lame and wanting. He said we also understand that the Earth does not readily surrender those things taken from her. The myths tell us so. Oedipus killed his father and married his mother. When she learned the truth, she hanged herself and Oedipus took two pins and blinded himself.

The ancient Greeks also told each other about Hephaestus, the god of the blacksmith's forge. Hephaestus was and is acknowledged, if not worshipped, by craftsmen of all kinds. He was the born son of Zeus and Hera, king and queen of the gods, and one day when he heard them having a furious argument he sought to intervene on his mother's behalf. Enraged, Zeus took Hephaestus by one leg and threw him from the summit of Mount Olympus.

His fall to Earth lasted a whole day and when he landed, on the island of Lemnos, he was broken and lamed. Another version has Hephaestus born to Hera who had coupled with neither Zeus nor any other, so he mattered less to the god than to the goddess. The parthenogenous birth was her revenge against Zeus for granting existence to Athena without her. Another version has Hephaestus born lame, with club feet. The shame of his imperfection had his mother throw him out of Olympus. In any event, Hephaestus was the smith of the gods, designing and making all their most famous weapons and attire – the winged helmet and sandals of Hermes, the armour of Achilles, the bow and arrows of Eros. His uncanny skills cost him dear.

A myth of northern Europe describes a wondrous smith called Wolund, or Wayland. His skills are coveted by a king, Nidud of Sweden, who sends men to seize him and bring him to his castle. Wayland is first hamstrung, so that he can never escape, then put to work at Nidud's forge. Always the smith is lamed, deformed or unfinished.

Just over a mile from the Uffington White Horse, in Oxford-shire, lies Wayland's Smithy. Thousands of years before the arrival in England of Anglo-Saxons and their version of the Wayland legend, Neolithic farmers raised a tomb for their dead. Sometime around the middle of the fourth millennium before the birth of Christ a timber chamber was built, more of a lidded box, really, into which were placed mortal remains. Archaeologists found one figure laid down intact, crouched and curled, like a baby in the womb, ready to be born anew. Nearby were the bones of eleven men, two women and a child. These latter had been stacked like lumber close to traces suggestive of a raised platform upon which their bodies might have been laid out, exposed to the elements and scavengers, so that their flesh was stripped prior to the placing in the box of those sticky remnants. There were arrow heads among the bones so some of the dead might have been victims of war. A pair of post holes at either end of the box may have been evidence of supports for a roof above all.

As with all such ancient tombs, it is impossible to be sure of much about intentions. Were other loved bodies placed in the box for a while, then removed for burial elsewhere? Were the fourteen found by the archaeologists the only tenants of that place of death? After just a few years an oval-shaped mound was raised over the woodwork and bones, from ditches dug either side, burying them entirely so that all was quiet for half a lifetime.

Some decades later, other Stone Age farmers reworked the tomb. A stone-built passage with one chamber either side, forming a cross-shape, was enfolded within a long, wedge-shaped barrow – 80 feet long, 40 feet wide at the front and 20 feet wide at the rear. Most imposing of all, and by far, half a dozen sarsen slabs, each 10 feet high, were erected as a dramatic façade across the front, framing a low-lintelled entrance. Within were laid the scattered remains of seven adults. Stone arrowheads were close by three pelvic bones. As before, as many questions as answers were buried with them. Whoever was interred, and why, it seems all that activity – in timber, then in stone – happened during a single century or so of the thousands of years of Wayland's Smithy.

Long years passed. The time of wood and antler, bone and stone was replaced, augmented rather, by the knowledge of

At Wayland's Smithy . . . a god fallen on hard times

metal-making. Ways of dealing with the dead changed too, bodies burned and turned to cinders that might be stored in jars. The tomb remained abandoned, all but forgotten, overlooked as a stony hump among beech trees.

During the fifth century AD, those Angles and Saxons came to the long island of Britain. Some arrived in the territory that would be Oxfordshire, likely via the ancient track called the Ridgeway, high ground linking west and east that has been walked over by our kind for five millennia at least and certainly longer.

Among much else the Ridgeway passes by the Uffington White Horse, the oldest geoglyph in the country. Cut into the face of the ridge three thousand years ago, it reminds me more of one of my wolfhounds in full gallop, but according to ancient consensus,

193

horse it is. Whatever the species, it is all sinuous movement and quite unlike any other of the white horses of England in terms of artistic brio. In Terry Pratchett's *A Hat Full of Sky*, the heroine, Tiffany Aching, wears a necklace bearing its likeness. Her father tells her: "'Tain't what a horse looks like. It's what a horse be.'"

The White Horse is close by Wayland's Smithy and perhaps that giant equine – 360 feet long from nose to tail – was in Anglo-Saxon minds when they encountered the ruined stones and made them think of warriors on horseback, and so a blacksmith and his weapon-making forge. It appears the looming façade, the cramped dark, the approximation of a cave into the underworld, persuaded some of those incomers that once upon a time the place had been home to their smith-god and his fires. Iron-work was found inside – a pair of unshaped bars that might have been blanks for swords – and folklore has always insisted on the presence there of the lame smith of the gods. Wayland is mentioned in *Beowulf*, the Old English epic poem about the eponymous hero's efforts to defend a clan of Danesmen against a monster called Grendel. As Beowulf prepares to fight, he leaves instructions regarding his mail shirt: 'If the battle takes me, send back this breast-webbing that Weland fashioned . . . to Lord Hygelac.'

The oldest mention of Wayland's Smithy is in a document prepared for King Eadred, in 955 A D, wherein 'Welandes Smithian' serves as a boundary marker between two territories. Surely those Anglo-Saxons had no way of discerning the truth of the site, but perhaps they saw human remains inside and made sense of them in the context of that broken god as he was understood by the peoples of northern Europe: a vengeful god . . .

A Norse poem describes Wayland's revenge on his captor.

Having lured King Nidud's sons to his forge, he kills and decapitates them, and from their remains sets about making grisly gifts for the royal family:

> I smote off the heads
> Of both thy sons . . .
> Their skulls, once hid
> By their hair, I took,
> Set them in silver
> And sent them to Nidud.
> Gems full fair
> From their eyes fashioned
> To Nidud's wife
> So wise I gave them.
> And from the teeth
> Of the twain I wrought
> A brooch for the breast
> To Bothvild I gave it . . .

Bothvild was Nidud's daughter and later Wayland lured her too, and drugged then raped her before flying to Nidud, using a magical cloak, so that he might gloat about his deeds.

Across the millennia, myths meet and mix, like mist, until one is hard to see in isolation from the others. Wayland seems to have ties to the tales of the Greek craftsman and builder Daedalus – the same who fled captivity on Crete, with his son, Icarus, using wings of his own devising. Icarus flew too close to the sun, of course, so the wax affixing the feathers of his wings melted and he fell to his death. Daedalus is credited with building the labyrinth beneath the palace at Knossos, home to the monstrous

Minotaur – and it has been suggested the Anglo-Saxons took one look at the passage and chambers of Wayland's Smithy and saw in them another maze.

Wayland's Smithy is haunted still by the ruined god, but by now he has taken to helping passers-by. Oxford antiquarian Francis Wise wrote, in a letter to one Dr Mead, published in *Concerning Antiquities in Berkshire*, in 1738: 'At this place lived formerly an invisible smith, and if a traveller's horse had lost a shoe upon the road, had no more to do than bring the horse to this place with a piece of money, and leaving both there for some little time, he might come and again and find the money gone, but the horse new shod.'

Novelist Sir Walter Scott grievously and utterly traduced the folklore in his novel *Kenilworth*, having his central character Tressilian, a Cornish knight, encounter there a 'strange blacksmith', then reveal him as no god, or any supernatural figure, but merely a mortal man eking a living by deception and a carefully hidden underground forge. In Rudyard Kipling's *Puck of Pook's Hill*, the eponymous central character is 'the oldest Old Thing in England'. He is beyond ancient and represents what the land once was: 'The People of the Hills have all left. I saw them come into Old England and I saw them go.'

Ancient or not, even Puck is an incomer to England. For all that, he was there to see Stonehenge newly built. The children to whom he addresses his stories are the descendants of more recent arrivals – Phoenicians and Gauls, Angles, Danes and Jutes. Puck describes how those immigrants brought their gods and notes that 'England is a bad country for gods – most of whom could not stand our climate.'

Wayland was one of those itinerant deities, but time and tribulation laid him low until, rather than a god demanding sacrifices upon the altar made of the tomb's lintel, he was reduced to scraping by on single coins in return for shoeing passing nags.

According to the locals, Wayland's Smithy is still a place best avoided at night. From time to time there are reports of the ringing out of blows struck with a hammer made of no earthly metal but of a stone that fell from the Moon. Visitors to the tomb report feelings of great, oppressive weight descending upon their shoulders, as well as sadness. Others claim a sensation of being watched, while those who venture inside the tomb report knowing at once they are not wanted there. More folklore has it that every hundred years the Uffington White Horse slips Earth's surly bonds and gallops the mile or so to Wayland's Smithy for fresh shoes.

I have wondered if it is the Old England of Puck that is there to be mourned not just at Wayland's Smithy but all along the Ridgeway.

Soon after the turn of the twentieth century, Anglo-French writer and historian Hilaire Belloc published *The Old Road*. He wrote about the road – the path or track trailing across the landscape from somewhere to elsewhere – as the most important and foundational of our species' impacts upon our world. Other creations were profoundly important too: 'We craved these things – the camp, the refuge, the sentinels in the dark, the hearth – before we made them; they are part of our human manner and when this civilization has perished they will reappear.'

But of the road he wrote:

> . . . it is the humblest and the most subtle but . . . the greatest and most original of the spells which we inherit from the earliest pioneers of our race. It was the most imperative and the first of our necessities. It is older than building and than wells; before we were quite men we knew it, for the animals still have it today; they seek their food and their drinking places, and, as I believe, their assemblies, by known tracks which they have made.

Folk still walk the Ridgeway today, stitching together, step by step, its surviving fragments. The White Horse is still there, so too Wayland's Smithy, of course. Kipling's Puck lamented the passing of the first settlers of the landscape: 'The People of the Hills have all left. I saw them come into Old England and I saw them go.'

Those departed figures represent the old way of life, the first way of life, and it is their ghosts that walkers of the Ridgeway might sense now. Whether we know it, or acknowledge it, we grieve for them, and harbour guilt that we made their world unliveable, drove them from 'the world that they were robbed of in their quiet paradise'.

Old England, Britain, is changing more and more, faster and faster. We know we have not done right by the old place and the guilt of it is a haunting presence, a grey figure at the corners of our eyes.

At the end of *The Old Road*, Belloc enters Canterbury Cathedral on the anniversary of the martyrdom of Thomas Becket: 'I had so fixed the date of this journey, the hour and the day were the day and hour of the murder.' He hopes that such an 'exact coincidence' might conspire with the significance of the place to grant him a glimpse of the archbishop as he had appeared that day of days: 'watching the door from the cloister, watching it unbarred at his command'.

Belloc might, he thought, have borne witness, across the abyss of centuries, to the crime itself re-enacted: 'the jangle of arms, and of scabbards trailing ... the sharp insults, the blows ...' Instead, he saw nothing but the empty church and any deeper experience was the stuff only of his imagination. 'There was no such vision,' he wrote. 'It seems that to an emptiness so utter not even ghosts can return.'

17

THE ANGEL OF MONS

'In the main, the result of our enquiry is negative, at least regards the question of whether any apparitions were seen on the battlefield, either at Mons or elsewhere. Of first-hand testimony we have received none at all, and of testimony at second-hand we have none that would justify us in assuming the occurrence of any supernormal phenomenon.'

SOCIETY FOR PSYCHICAL RESEARCH

'Then there is the story of the "Angels of Mons" going around the 2nd Corps, of how the angel of the Lord on the traditional white horse, and clad all in white with flaming sword, faced the advancing Germans at Mons and forbade them further progress. Men's nerves and imaginations play weird pranks in these strenuous times. All the same the angel at Mons interests me. I cannot find how the legend arose.'

BRIGADIER GENERAL JOHN CHARTERIS, AT GHQ

HAUNTINGS

'Every blade of grass has its angel that bends over it whispering "grow, grow".'

TALMUD

MORE THAN GHOSTS HAUNT US. DAD ONCE TOLD US ABOUT SEEING AN ANGEL BY my mum's bedside. It was years ago – so long that I may still have been living at home at the time. I cannot remember for certain when it happened. What I do recall, and clearly, is his account. He had awoken in the night needing to answer a call of nature. On his return, he stopped abruptly in the bedroom doorway. In the darkness within, softly illuminated by the light from the hallway, he saw a tall figure, taller than a man, standing by the far side of the bed – Mum's side – and looking down at her while she slept. The figure was dark and hooded, somewhat indistinct, but its presence, my dad was sure of it, was benevolent and caring. He knew it for a fact. Dad said it paid him no heed, just continued to watch over my mum. After a few moments, the apparition vanished, dissolved into the surrounding darkness. As far as Dad was concerned, he had seen Mum's guardian angel.

What do I know? Not enough – I know that much. That is what my dad said he saw, and I took him at his word, then and now. It was the middle of the night, and a mind still floating on the surface of sleep might play tricks on a person. He made no song and dance about his vision – just recounted it when asked. Over the years I asked him about it, and he would only repeat what he had said the first time. The matter-of-factness of his testimony convinced me only, and always, that he truly believed he had seen what he had seen.

*

On 30 August 1914 *The Sunday Times* carried news of a devastating reverse for the British Army, a week before, at the Battle of Mons. They had been routed, that much is certain, but all, however, may not have been exactly as it was portrayed in the newspaper. Outnumbered by the enemy though they were, the British force had had at its core men who had fought in the Boer War and had learned from that South African experience all about quickly digging defensive positions and exploiting the shapes and flows of landscape for cover. Furthermore, they had acquired a reputation for rapid, accurate fire with their Lee Enfield rifles such that their enemies often thought their foe must be using machineguns. All in all, they were well suited to the kind of open, fast-moving events that unfolded at Mons and in its aftermath.

When their French allies nearby began a speedy retreat in the face of overwhelming numbers, the men of the British Expeditionary Force were well able to conduct a fighting retreat. On the day, they suffered around sixteen hundred casualties compared to more than five thousand for the Germans. It seems the *Sunday Times* piece therefore painted a deliberately bleak picture: 'We have to face the fact that the British Expeditionary Force, which bore the great weight of the blow, has suffered terrible losses and requires immediate and immense reinforcement ... [it] needs men, men and yet more men.'

The objective of the piece was less accurate reporting and more a well-taken opportunity to call for volunteers to join the war effort.

It was in the immediate aftermath that Welsh author Arthur Machen penned a short story called 'The Bowmen', about a

British soldier at Mons crying out for help from St George and thereby summoning a host of phantom archers. Those bowmen shot hails of arrows towards the foe, and while the Germans fell dead by the hundreds, their bodies were left unmarked. The supernatural intervention gave the British force vital time to pull back to safety – and a legend was born.

Even though 'The Bowmen' was purely a work of fiction – appearing first in the London *Evening News* on 29 September – it captured imaginations. By 1915, newspapers and magazines were receiving letters claiming to provide testimony from soldiers and others who had witnessed the ghostly archers during the real battle. Crucially, none was a named first-hand account; they had always been heard at second or third or fourth hand from some-one else, some soldier or officer or nurse who had been there that day. As well as bowmen, there was soon talk, too, of angels.

In an attempt to stem the flow of myth-making, Machen wrote an introduction to an anthology published that year – *The Bowmen and Other Legends of the War* – in which he sought to make clear his story was mere fantasy. To no avail. To the end of his days, Machen would continue to reiterate that 'The Bowmen' was a product only of his imagination. But a British population in need of hope preferred not to let inconvenient facts get in the way of a good story, and long after the war, to this day, the story of angels and fighting men appearing in the hour of need at Mons has its faithful adherents.

Like nacre laid down by an oyster, the massed myth-making formed a pearl around the grit of Machen's tale until the facts of it – or, rather, the fiction – was utterly concealed. It was no longer his story. Instead it had been taken up by its audience, possessed and made their own, made into something that was real – at least to them.

One 'eyewitness' after another reported seeing a great figure clad in white, upon a white horse, wielding a sword of dazzling brightness. Others reported ghostly horsemen upon phantom horses, escorting British soldiers to safety or otherwise covering their retreat and preventing the Germans getting close, rendering their weapons harmless. William George Ludlow was seventeen years old at the Battle of Mons. Long after the war he was still telling his children and grandchildren about having watched a beautiful angel single-handedly galling the German advance.

In an article in the *Eastern Daily Press* on 25 June 2017,

When hope still dwelt in soldiers' hearts . . . a heavenly host defies the foe

William's grandson, David Ludlow, recalled a Christmas when his grandfather talked to him in detail about what he had seen: 'When I was asking what the angels had been like, he pointed to a card on the mantelpiece with a picture of an angel with wings outspread and said, "It was just like that." '

Bill Ludlow survived Mons, Ypres, the Somme and more. After the war he picked up work as a labourer at a cement factory. He married but his wife died of tuberculosis soon after the birth of their son, also William, David's father. David had a close relationship with his grandfather and heard him talk about Mons more than once. 'My grandfather said he saw this angel . . . twenty feet tall with outspread wings, hands behind her, holding back the [German] lines. He could see her face plainly – beautiful, he said.'

The diehards of the BEF at Mons were long enough in the tooth to have heard more than one story of old battles, old wars. Many would have known the story of the turning of the tide at Agincourt by archers who darkened the sky with their arrows. Did that make them more or less susceptible to the legend? And what of a callow boy like William? Was the fear of combat, coupled with exhaustion born of sleep-deprived nights, enough to have him hallucinate?

An anonymous letter to the Bath Society paper in 1915 was typical of those who reported sighting not archers but angels fitting the description familiar to men of traditional Christian upbringing: 'I myself saw the angels who saved our left wing from the Germans during the retreat from Mons. We heard the German cavalry tearing after us and ran for a place where we thought a stand could be made. We saw between us and the enemy a whole troop of angels.'

A recurrent theme had the angels of Mons, of whatever form, enfolded within something opalescent as a pearl – a bright, glowing cloud, or a beguiling mist within which luminous figures moved.

(In a strange and dreadful inversion of the image of the life-saving cloud of angels, 22 April 1915 saw the first use of poison gas, at Ypres. At around 5 p.m. cylinders buried in shallow graves by specialist German soldiers began to release their contents. A greenish-yellow fog rose like a ghost and drifted on the wind towards trenches held by French soldiers, specifically colonial recruits from Algeria. The effect of the chlorine cloud was witnessed by a British officer: 'A panic-stricken rabble of Turcos and Zouaves with grey faces and protruding eyeballs, clutching their throats and choking as they ran, many of them dropping in their tracks and lying on the sodden earth with limbs convulsed and features distorted in death.'

If no divine forms had come to Mons to fight for life, the angel of death, conjured by mankind, had risen from the ground at Ypres. A British population fearful for the fates of fathers and sons, and desperate for any good news, was primed to accept any suggestion that God was on their side against a godless enemy. It was in an atmosphere of man-made horror and – with stories of ugly death all around and so much more to come – that the legend of the Angel of Mons was summoned into being, by and for anxious and frightened British hearts.)

Or was Arthur Machen wrong in assuming, insisting that it was his story and only his story that sowed the seeds of myth-making? It seems hard to accept that so many men – and women of the Nursing Corps – would either lie or be mistaken in the same way at the same time. Professor of clinical psychology Mattias Desmet, of Ghent University in Belgium, has written about mass formation psychosis, how 'free-floating' anxiety within a group already stressed by overwhelming circumstances they do not understand, and which frighten them, may make them vulnerable to a compelling explanation suggested to them by someone they trust. In this way, according to Desmet, many thousands of people may come to share the same unifying notion at the same time. Were men worn down by battle, fatigued and hungry after sustained effort in a grand life-threatening situation, susceptible to the suggestion of divine intervention? The numbers were against them after all – and disaster seemed imminent – yet from the jaws of defeat was snatched an escape that seemed inexplicable, almost miraculous. Perhaps some of those exhausted and strung-out men were subject to hallucinations, and when they talked about what they had seen found others who had endured similar experiences and wanted the reassurance that they were not 'mad'.

Whatever did or did not happen at Mons in August 1914, it was by no means the first time men *in extremis* had claimed, even perhaps believed, they had been saved by angels. No less a warrior than George Washington said his actions at Valley Forge, between December 1777 and June 1778, were directed by angels – and in the aftermath of the bloodbath of Shiloh, 6–7 April 1862, soldiers said they were guided to the locations of injured men by shining clouds.

Having captured the attention of a population in the grip of war, the Angels of Mons never really let go. The embers were always glowing, awaiting fresh oxygen. The episode was fanned back into life in 1999 when a photograph surfaced in a Welsh antiques shop. Danny Sullivan, an author from Stroud, told the press he had stumbled upon a canister of film and other items in Bonita's shop in Agincourt Square in Monmouth. He told a journalist at *The Sunday Times* that, according to accompanying letters and other documents, the film had been shot by a Mons veteran named William Doidge, also from Monmouth. Having seen the ghostly host with his own eyes during the battle, while serving with the Scots Guards, Doidge had developed a lifelong obsession with the paranormal. In an added twist, he had written about how he had fallen in love with a local girl while on active service on the Western Front. Having lost contact with her by the end of the war, he had grown convinced the Angel of Mons could reunite them.

In 1952 his path crossed with that of an American First World War veteran, identified only by his first name, Doug, who had his own story to tell. According to Sullivan, Doug told Doidge about a tragedy that had befallen some American soldiers training for D-Day in Woodchester Park, the grounds of a Gothic Revival pile near Nympsfield in Gloucestershire. A pontoon bridge across a lake had collapsed, crushing and drowning several men. According to Doidge's GI, an angel, like that reported by witnesses at Mons, had been seen hovering over the lake the day before the accident. Along with the canister of film, and paperwork collected by Doidge, there was a black-and-white photograph he had taken of the angel in question, which circulates still on the internet.

The rest of the world's media got involved when Sullivan claimed he had sold the film and the story to director Tony Kaye (behind, among other films, *American History X*) for half a million dollars. According to Sullivan, Marlon Brando had been signed to play the part of Doidge in a blockbuster Hollywood movie.

In a piece published in the *Guardian* on 13 March 2001, Kaye was reported as saying the epic would bring together the drama of *Titanic* with the otherworldly spookiness of *The Blair Witch Project*. Commenting on the contents of the canister from the antiques shop he added: 'This is the closest we have on film to proof of an angel. I've spent much of my life looking at special visual effects, and this is an effect for which I have no explanation.'

So far, so supernatural, until 2002 when investigation by Chris Morris, a journalist working for BBC Wales, found the whole thing was a hoax. Sullivan had concocted the story – William Doidge, Doug the American GI, the angel at the lake, the existence of the film and all the documents, the movie deal – as part of a scam to draw attention to a book he had written a decade before about the haunted history of Woodchester Mansion. Tony Kaye, the film director, had been in on the ruse as well.

As might be expected, the Angel of Mons will not die. By now there are innumerable books specifically about or making mention of those alleged events of August 1914, those that keep faith with the legend and those that take Machen's line and insist the whole affair was fiction, which enthralled and persuaded those in need of comfort and reassurance that good and God will prevail in evil times. The William Doidge hoax notwithstanding, the legend is referenced, too, in more than one movie so it is handed down from generation to generation, the separation between imagination and reality never clear.

The notion of divine intervention by shining figures wrapped in light was there before the First World War and gave service after. An online search for Padre Pio will generate entries detailing the many miracles attributed by the Catholic faithful to a Capuchin friar and priest famed for bearing stigmata – the bloodied wounds suffered by Christ – for most of his life. Multiple sites describe how whenever US bombers were sent anywhere near San Giovanni Rotondo, Padre Pio's hometown, they were diverted from their destructive paths by the appearance in the sky ahead of the giant shining image of the friar, arms outstretched in the universal gesture of peace. The *Washington Post*, on 29 November 1983, carried a review of a book about Padre Pio by English author Suzanne St Albans, in which journalist Phil McCombs included the miracle of the redirected bombers:

> During the Second World War, according to Father Joseph, who had been the padre's top aide, 'American bombers frequently flew in this direction . . . But it happened more than once that pilots disobeyed their orders and returned to base without having got rid of their load. The reason for this is given in the records as: "The vision of a monk standing in the sky, diverting aircraft."'

That the First World War population was so in need of the thought of angels protecting their men and boys is easy enough to understand. And that those soldiers who took part in the first grand battle of 1914 were ready to be persuaded of a miracle is also readily to be forgiven. Their 'miracle' came early in a war that was shortly to be transformed, by trenches and machine-guns, into something almost infinitely worse than anything experienced at Mons.

HAUNTINGS

By dying before the horror of the Western Front, Rupert Brooke had preserved innocence – his own and innocence itself – completely and neatly. He had, still, the young man's appetite for war of the sort he had surely read about in Homer, fought by clean-limbed heroes, which was wondered at by A. E. Housman's 'lightfoot lads' so that the clash of arms was still to be desired . . .

To turn, as swimmers into cleanness leaping,
Glad from a world grown old and cold and weary.

The soldiers at Mons may have caught a glimpse of the future, its awful enormity. But perhaps for them in those moments there was, too, still room in hearts and minds for shining hope. In such unviolated spaces, it was still possible to conceive of shining angels, spectral bowmen sent by St George. Brooke had also had, however, the premonition of the reality to come, aboard his ship en route to Gallipoli in the days before he died . . .

Pride in their strength and in the weight and firmness
And link'd beauty of bodies, and pity that
This gay machine of splendour 'ld soon be broken
Thought little of, pashed, scattered . . .
Like coloured shadows, thinner than filmy glass . . .
Perishing things and strange ghosts – soon to die
To other ghosts – this one, or that, or I.

It was Ypres that was coming, and Arras and the Somme and Passchendaele. At Mons, so early in the war, perhaps, in the minds of experienced fighting men, who could read the runes and sense that something altogether different was coming, arose the presentiment of it all. In the years ahead, when physical and spiritual exhaustion meant survival was the very most that might be hoped for, there would be no room for fantasies of divinely wrought salvation. No more heroes then, real or imagined – no Arthurs, no saints, no bowmen of Agincourt.

And yet, and yet . . . That the story lingered, has lingered, like a fire sent underground, smouldering in a buried seam of coal, only waiting for reignition on contact with the air, surely makes plain our human need for spirits from realms invisible. We are twenty-first-century people yet still we hanker after ancient notions. Much later, Machen said of the indomitability of his imagining: 'It was as if I had touched the button and set in action a terrific, complicated mechanism of rumours that pretended to be sworn truth, of gossip that posed as evidence, of wild tarradiddles that good men most firmly believed.'

My dad's dad, my grandfather Robert Miller Oliver, was at the Somme and Passchendaele and lived . . . lived *not* to tell the tale, as it turned out, silent as he was about whatever had happened to him there and then. I have wondered if my dad was imbued, epigenetically, with some of its effects by way of inherited DNA. When I think now about what happened to him that night, when he said he saw my mother's guardian angel, I make space in my head for the thought that what he had then was a gloomy premonition of his own death . . . of his dying first and leaving my mum behind. And with that inkling, befuddled by sleep, it was his own self he saw, watching over her while she slept, not knowing he was there.

18

LUD'S CHURCH AND A LOLLARD MARTYR CALLED ALICE

'My son, if thou are of the Protestant religion and are called upon by the despotic rulers of this land to abjure thy faith, remember the Lollards of Ludchurch and stand firm.'

ANONYMOUS

LUD'S CHURCH IS A DARK WOUND SLICED FATHOMS DEEP INTO A SLAB OF MILLSTONE grit close by the village of Flash in Derbyshire. A landslip long ago, a parting of the ways, split a great mass of rock in twain, and made a cleft as narrow as an alleyway between two city blocks. More than a hundred yards long and sixty feet deep, it has been a furtive, sinister hidey-hole for who knows how long. Tens of thousands of years. Looming above and beyond are the cracked and jagged outcrops that shape the character of Derbyshire and North Staffordshire: Gradbach Hill, Ramshaw Rocks, the Roaches – the 'rogh knokled knarres with knorned stones', as the author of *Sir Gawain and the Green Knight* has it – names as tough-sounding as the millstone grit itself, so named on account of its suitability for millers' grindstones.

Up there the sun shines, grouse call and the wind blows on heather moors, wild and wuthering. But in the perpetual gloom of Lud's Church warmth is a rumour, scarcely to be believed. Sometimes that slit makes its presence felt in the countryside nearby with cold air that issues from it, like a deathly exhalation. Life finds a way, as they say, and its chill sides are home to those plants that thrive where others won't – ferns, silver hair grass and liverwort.

Scholars of literature have suggested Lud's Church might be the Green Chapel of Sir Gawain and the Green Knight. Penned by a

Rock of Ages cleft . . . the chapel of the Green Knight

nameless poet, it tells the tale of King Arthur's nephew and youngest knight, Gawain, who accepts a challenge from a mysterious figure – green of flesh, of garb and mounted on a green horse. Having arrived at Camelot on New Year's Eve, the Green Knight offers to play the beheading game: he will accept a blow from an axe if the axeman will visit him at his home – the Green Chapel – in a year and a day. Gawain accepts and duly decapitates the uninvited guest. Utterly unconcerned, the Knight picks up his head, bids the company farewell and leaves. Gawain embarks upon his quest, having his vows of chivalry and chastity tested along the way before finally meeting the Green Knight in what is essentially a great mound with a long passage inside it – something of a Wayland's Smithy, in fact, but on a colossal scale.

> . . . aboute hit he walkez
> Debatande with himself quat hit be mgyht
> Hit had a hole on the ende and on anyther side
> And overgrowen with gres in glodes anywhere
> And all was holw inwith, nobot an olde cave,
> or a crevisse of an olde cragge . . .

Sir Gawain and the Green Knight is written in Middle English, in a Midlands dialect philologists have placed in the vicinity of Derbyshire, Cheshire and Staffordshire, encouraging academics to concede the likelihood that the anonymous poet hailed from thereabouts and may have been inspired by Lud's Church. The eponymous knight bears the colour of life itself, shading him with the tones of Jack in the Green and therefore the mythology of the living spirit of the wood. On the day of the equinox in medieval times, the youth of the land would cover themselves in mosses and other natural greenery to make manifest the 'Moss Man'.

But green is, too, the colour of evil, made plain by countless references – Shakespeare's green-eyed monster of jealousy in *Othello* . . . green with envy. The Devil was associated with green – so, too, his servants, witches and the like. Green was the imagined colour of serpents and dragons, and during the Crusades, it took on yet more associations with the enemy, the adversary, when it was learned that Muhammad's favourite colour was green. Green was also faithless – most likely of all tones to fade and discolour in the hands of dyers – and therefore emblematic of all that was unpredictable in the cosmos, changeable and untrustworthy.

Anyone descending from the day, from a hillside above the Dane river and deeply down into the gloom, might be forgiven for sensing ill-will in Lud's Church. The cleansing of sunlight is absent, and since trees on the heights above, oak and rowan, have conspired to make a roof over much of the winding length, the blue of the sky is seldom to be glimpsed. Passers-by leave trinkets, mostly coins in split wood or other crannies, echoes of pagan rites of sacrifice and mollification of gods. There are stories of Druids worshipping there long before Lud's Church was a church.

The name Lud might be attached to one Walter de Lud-Auk, a local Lollard – Lollardy being a sect with its origins in the teachings of the fourteenth-century English theologian and philosopher John Wycliffe. His principal sin in the eyes of the Church was his commitment to translating the Bible from Latin into the everyday language of the people, and he also questioned many of the ways of the Church, such as the worship of saints, the legitimacy of the papacy, the celibacy of priests and the belief that bread and wine might actually be transformed into the flesh and blood of Jesus. Since his thinking was heresy, his followers were ever after condemned and persecuted. The name Lollard was not one they had taken for themselves. Rather, it began as a pejorative, derived from a Dutch word for the benighted brotherhood that had buried the dead of the Black Death, and chanted softly while they did so. Since *lollen* was the verb 'to mutter', Lollard was a corruption applied to those who quoted holy scripture in the common tongue.

Swythamley and Its Neighbourhood, Past and Present is a nineteenth-century guidebook by local landowner Sir Philip Brocklehurst, which recounts, among other stories, what happened to Walter de Lud-Auk (or perhaps Ludank) and a party of Lollards, including Alice, his beautiful eighteen-year-old niece, on some lost summer afternoon. They numbered fourteen in all, protected at all times by Henry Montair, keeper of the woodland that provided so much seclusion for the ravine. The Lollard faith was outcast, and Lollards met not in churches but without, always seeking privacy. During the summer months, when the weather was kinder, Walter and his followers took to living in that place they called Ludchurch. Among his services to them, Henry would keep the little congregation provided with food and other supplies.

Years passed, Ludchurch keeping them safe in all weathers. By the time of the dread day, Walter was in his seventies, white-haired but still vigorous and strong. Alice was his ward, her parents having died, and she was always by his side. She had a fine singing voice and those pure notes that had been the congregation's blessing would finally be their curse. The ravine had been a fearful place before the coming of the Lollards. They had brought something good and clean, uplifting, but the deeper nature of the place was about to reassert itself and a place of evil it would be again. First Walter led them in prayer, then had them lift their voices in song. Perhaps sensing danger nearby, one by one they fell silent – until the only voice was that of Alice, oblivious. Her singing cut through the gloom all about, and it was their undoing. Soldiers nearby, ever on the hunt for heretics, heard the hymn and rushed into the ravine.

'Yield in the name of the Blessed Church,' shouted their commander.

Even non-attendance at official church services was regarded as a grievous moral wrong, while finding alternative ways to worship was strictly illegal. Lollardy and other fledgling protest took place before the Reformation but John Wycliffe has been cited by some as the bright star that heralded its advent.

Some of Walter's men reached reflexively for their weapons but he told them no. Riled beyond endurance by the rough manner of the soldiers, however, Henry Montair, a huge man with strength to match, grabbed one and flung him against the scrum of his fellows, like a bowling ball at skittles. It was then, in the ensuing chaos, that one of the soldiers levelled his musket, likely a primitive piece called an arquebus, and fired at Montair. The shot missed the forester but found its mark instead in the breast of Alice de Lud-Auk, standing just behind. She died in her grandfather's arms and Montair, in his horror and rage, set about the uniformed men, laying down one after another. Single-handedly he drove them back towards the entrance to Ludchurch and was stopped only by the awful, mournful sound of a lament behind him. For the last time, those Lollards raised their voices to sing of the loss of one of their own, their finest flower.

Walter carried Alice out of the ravine, and, under the watchful eyes of the soldiers chastened by her death, she was laid to rest in a grave hastily cut. There were prayers then, before the whole group – Montair included – surrendered themselves into the custody of the soldiers. Montair would later escape, making his way to France, later returning to England to continue to fight against the Establishment. Today an oak tree near the mouth of Ludchurch is regarded by some as marking Alice's grave. Now among the looming shadows of the ravine they say it is Alice's ghost that sadly roams, Alice's voice that is heard carried on the wind. Around 1862, Brocklehurst installed, in a niche within the ravine, a carved wooden figurehead scavenged from the wreck of a ship named *Swythamley*. Time and the dankness have long since rotted it to nothing.

Despite the romance of the tragedy of Alice and Walter, and those Lollards, not all are persuaded that Lud's Church owes its name to the luckless de Lud-Auk family. An alternative etymology has deeper roots, reaching all the way to the mythical King Lud, named in Geoffrey of Monmouth's *History of the Kings of Britain* as the monarch who founded London and lies buried at Ludgate. The name of the gate, it is generally accepted, more likely comes from the Old English *hild-geat*, meaning 'swing-gate', but those faithful to King Lud claim it is he that is remembered in the naming of the ravine. Lud may be the Welsh hero Llud Llaw Ereint, meaning Llud of the Silver Hand: he lost his arm in battle and had it replaced with a glittering prosthetic crafted by his brother, Dian Cecht.

The stories at Lud's Church come so thick they must be batted away like clouds of swarming midges. But strip away all talk of Druids and Lollards – there is even talk of the place being the stamping ground of a boggart, a malevolent creature of English folklore, given to mischief and the abduction of children – and Lud's Church is still likely to capture the imagination. It is a wonder. Perhaps the only malevolent darkness is carried in by us, flawed human beings shaded by our regrets, wrongdoings and guilty consciences.

There are innumerable such locations, special places folded into the landscape of the British Isles and often with mystical associations. Finnich Glen, in Stirlingshire, is every bit as hidden as Lud's Church, and as captivating. Running east of Finnich Burn it is another work of nature – patient work, that is, rather than the product of strenuous upheaval like the landslip that made Lud's. The water of the Carnock Burn, perhaps pressurized at times by the weight of melting ice at the end of the last Ice Age, has cut a deep and almost claustrophobically narrow chasm into red sandstone. As at Lud's Church, the vertiginous walls, soaring high, are clothed green with mosses, lichens and other species adapted to perpetual damp and gloom. On account of the red rock, the water is stained as though with blood. An outcrop, shaped and worn until it resembles a mushroom, or perhaps a hat, is called the Devil's Pulpit (of course it is) where the Adversary himself was apparently wont to preach to his followers from time to time. Again, like Lud's Church, Finnich Glen is associated by some with the nature worship of Druids.

St Nectan's Glen, another emerald treasure, near Trethevy in Cornwall, is the work of the River Trevillet. A sixty-foot waterfall sings its silvery song there and, by the shallow pool it makes below, visitors leave trinkets – photographs of loved ones, letters, toys, crystals, coloured ribbons and the like. Others build tiny towers of flat stones (the sort that might be chosen for skipping) called fairy stacks. All of it is made by people in pain, and left in hopes of intercession either by the Christian saint who gives the place its name, or from the curative power of nature, cradled in a place of impossible loveliness.

Lud's Church has earned somehow a dark reputation. The death of Alice de Lud-Auk is at the heart of it but in every other way the ravine seems to me only a special offering, a trick of nature conjured only to please and confound the senses.

Walkers passing through the ravine have, from time to time, reported sightings of drifting lights, so-called 'ghost lights' – amorphous, luminous shapes appearing, moving and disappearing in the manner of will-o'-the-wisps. British author Paul Devereux has written extensively on his website about 'earth lights', a term he has coined for phenomena he attributes, in part, to 'stresses and strains in the earth's crust'. Tectonic Strain Theory posits the notion that seismic activity – even that which is otherwise undetectable by human senses – might manifest effects in the atmosphere above ground akin to ball lightning. Descriptions of lights in the sky are not new:

Earth lights are or were known to traditional and ancient peoples [Devereux wrote]. They were fairies to the Irish and other Celtic peoples, though also harbingers of death known as 'corpse candles' to the Welsh; the disembodied heads of women who had died in childbirth to Malaysians . . . Bodhisattvas to Chinese or Tibetan Buddhists, who built temples where such sightings occurred . . .

Since Lud's Church is described as having been created by a massive landslip in ancient times, it seems reasonable at least briefly to entertain the thought that ongoing geological activity, perceptible or not, might generate the kinds of floating lights reported there more than once.

Devereux has also described the phenomenon as 'ancestor lights', as though they might connect us with those who went before. Years ago, I went hunting for the ancestors of Vikings on the tiny island of Faro, off the northern tip of Gotland, in the Baltic Sea. Archaeologist Joakim Wehlin took me to look for four-thousand-year-old rock art – etchings of ships made by sea-going people long, long before any Scandinavian ever thought of going a-Viking.

We found the petroglyphs in question on a flat slab of bedrock, one of many outcrops exposed, like knees through worn trousers, on a swathe of moorland shadowed by low hills. It was late after-noon in winter and the sun was soon to disappear into the invisible. We had been running out of daylight for the necessary filming so our finding of the ephemeral traces, in the nick of time, was satisfying. The moorland was partially flooded, and the rock outcrop in question submerged beneath some inches of water as clear as crystal, topped with a skin of ice, like polished window glass. The etchings had been lightly wrought – a single dish-shaped line for each vessel and a dozen or so vertical strokes within, representing individual oarsmen. That they were separated from our pointing fingers by a crust of ice, like museum exhibits behind glass, only added to the distance, physical and temporal, keeping us from the artist. He and they were beyond our reach in every way.

Back in Lud's Church we know who's there or rumoured to be. Sir Gawain is fiction but the poet walked that way, for sure, in the years before he conjured his tale. He stopped and looked about him, smelt the same smells, glimpsed the same thin slice of sky above. Even the existence of Walter and Alice, her fate and his ever-lasting sadness, is less than certain. They are all ghosts, essences hanging in the air, like the damp. I am haunted by encounters like those, so many down the years. Glimpses of the intentions of the ancients, but no physical connection with them.

Lud's Church is haunted not just by poor doomed Alice, but by the ghosts of all those who passed that way. Ice Age hunters surely found that slot; what they made of it we cannot know. Was it haunting and haunted for them? And then all the rest: the first farmers who found their way into the uplands, then Druids, Lollards and all else in between. In that rock of ages, cleft for them, some were surely struck by the place, the residue of their emotions snagged, like down blown from a dandelion clock. Whenever any of us does likewise, we brush past it and feel its light touch.

19

St Peter's Church, Luddites and the Ghost in the Machine

'They said Ned Ludd was an idiot boy . . .
That all he could do was wreck and destroy . . .
He turned to his workmates and said: "Death to Machines" . . .
They tread on our future and they stamp on our dreams.'

ROBERT CLAVERT, 'NED LUDD'

AFTER LUD, LUDDITES. NO RELATION.

The Reverend Patrick Brontë, father of the literary sisters, was an assistant curate at St Peter's Church, in the village of Harts-head, West Yorkshire, between 1811 and 1815, momentous years taking in the War of 1812 with America and the Battle of Waterloo against the French. By nature a conservative man, Patrick was no supporter of the workers' unrest, violent and impassioned, that was rumbling around the north of England, like bad weather. That said, local folklore in the villages around his former church tells a haunting tale in which Patrick features, a watcher all but invisible in the shadows.

Years of strife with Napoleon had brought hardship to Britain – not just to the long-suffering poor but to those higher up the food chain as well. As historian Frank Peel wrote in *The Risings of the Luddites, Chartists and Pluggers*, published in 1888, 'The hard pinch of poverty was now felt in many a dwelling where it had hitherto been a stranger.' Men came back from those wars and found themselves redundant, even unwanted.

In the dark of evening on 11 April 1812, men gathered quietly, by prior arrangement, in the shadow of an obelisk at nearby Cooper Bridge called the Dumb Steeple. A squared column of local millstone topped with a stone ball, it marked the spot where two roads crossed – the one linking Huddersfield and Liversedge, the other joining Brighouse to Mirfield. When anyone thereabouts was asked why it was called 'dumb', the stock reply was that it had never been heard to speak. By the time they set out along the road across Hartshead Moor towards Rawfolds Mill, at Cleckheaton, some miles away, those men, speaking in low voices so as not to declare their presence overmuch, numbered perhaps 150. Peel wrote:

They were armed in very motley fashion: some bore guns, others pistols, while many carried only hedge stakes or stout bludgeons of various kinds, and not a few held on their shoulders huge hammers, mauls, and murderous looking hatchets of various sizes. They were nearly all disguised, some having their faces simply blackened and others wearing masks to conceal their features effectually. Many of them were dressed in carter's smock frocks, others had their coats turned inside out, some had put their checked shirts over their clothes, and a few had actually dressed themselves partly in women's apparel.

Rawfolds Mill was owned by one William Cartwright. He had installed automated looms, driven by a water wheel, and by so doing had cut skilled men, trained to work by hand, out of a job. He knew all too well how his machines enraged those who depended, for earning a living, on skills hard won, and had taken the precaution of summoning soldiers and other armed men to protect his property.

When the men from Hartshead arrived, there was a brief but bloody fight. Shots were fired by both sides, but the defenders succeeded in driving off the attackers. Wounded men were shepherded away by their compatriots, succumbing later to their wounds. The story told at Hartshead is that graves were hastily dug in consecrated ground by St Peter's Church and men laid down without stones to mark their places. Patrick Brontë is said to have witnessed as much, and while he neither helped nor blessed the burials, he said no more about it after the turf had been stamped down. To this day a plot of ground is left undisturbed in the belief it holds those unmarked graves.

For two hundred years the White Gate Inn has sat beside Leeds Road. Older yet is the Three Nuns, at nearby Mirfield. Both were gathering points for men bent on wrecking Rawfolds and might have offered comfort to the wounded and dying on their return. The White Gate Inn is still on its original footprint, extensively renovated though it has been over the years. The present pub on the site of the Three Nuns was built in 1939, but buried under the car park are the foundations of a guesthouse and hostelry razed in the fifteenth century. All manner of stories are told about the Three Nuns – and its name is testament to its former owners, Katherine Grice, Joan Leverthorpe and Cecilia Topcliffe. The trio were the last of the nuns made homeless by the dissolution of Kirklees Priory during the reign of Henry VIII.

According to tradition, Katherine was wooed, but not wed, by one of Henry's commissioners. Finding she was pregnant, and abandoned by the father, she drowned herself in a nearby stream known now as Nunsbrook. In 1644, en route to the Battle of Marston Moor, Oliver Cromwell spent the night there too. Lost beneath the cars or not, at the Three Nuns it's history all the way down. The pub is associated with the prioress – her name lost to history – who, when asked to treat the wounded Robin Hood, bled him to death and ordered him buried by the side of the nearest road.

Regulars at the Three Nuns and the White Gate Inn have reported encounters with the wraiths of some of those who died that night in April 1812. The usual accounts of cold spots, mysterious footsteps, doors opening and closing, bolting and unbolting, are attributed at least in part to those unquiet dead.

They called themselves Luddites, the angry men who set themselves to reclaiming from new-fangled technology their right to work. There is a story that resonates with us in a twenty-first century bedevilled by suspicions about artificial intelligence (AI) and the rest of the digital tech likely to consign whole new generations to redundancy. For a long time we've been encouraged to laugh at the mention of Luddites, a label applied to any who rage against the machines. Now the last laugh, hollow from the chests of long-dead Luddites, may yet be on us.

In the time of those original Luddites, much of the mechanizing of industry had been under way for a long time. During the reign of Queen Elizabeth I, English clergyman and inventor William Lee of Calverton had come up with a mechanical knitting machine – the stocking frame. Good Queen Bess was sympathetic to the plight of her working people and refused him a patent on the grounds it would take work away from craftsmen. James VI and I was similarly protectionist. Just as toothpaste does not go back into the tube, though, Lee's technology stuck around. Machinery took jobs from time to time and that was that; people learned to live with it. But from the beginning, there were those who took against the machines.

While unskilled labourers had been growing used to being herded into mills to service machines, those with hard-earned skills were more defiant. After seven-year apprenticeships, men earning decent money to hand-finish fabrics, the so-called 'croppers', felt more than entitled to jobs, and wages commensurate with their skills. Rising first in Nottingham, then West Yorkshire and Lancashire, then Derbyshire and Leicestershire, croppers and others set about smashing stocking frames and other machines.

They took their name and inspiration from a near-mythical figure called Ned Ludd, or perhaps Ludlam, said to have been a young Leicester weaver. For those convinced of his existence, a day had come when Ned had taken furious umbrage at his stocking frame and smashed it, then another, with a hammer. That had been in 1779 and from then on, whenever frames or suchlike were found wrecked, the joke was that 'Ned Ludd did it.'

Anger simmered. On 11 March 1811 a gathering of men and women protesting about their plight in the face of industrialization was violently dispersed by a company of soldiers under orders from the Crown. In response, a mill in Arnold, Nottinghamshire, was damaged, and from there the movement spread. There were deaths on both sides. In April 1812 thousands protested at a mill in Manchester. Defenders fired their muskets into the crowd, killing three and wounding eighteen. Elsewhere, mill owner William Horsfall let it be known he would 'ride up to his saddle in Luddite Blood'. During the subsequent defence of one of his properties in West Yorkshire, he was shot and killed. Three Luddites were hanged for the crime. The thunderclouds of rebellion rolled and rumbled for years, with countless protests and incidents of damage. All across the north of England, men gathered by night and set about the destruction of the hated machinery. Letters warning of their intentions and demands were signed Ned Ludd, or Captain, General or King Ludd. In a knowing nod to the legend of Robin Hood, who had bled out in the priory close to the sites of those haunted pubs, they gave his location as Sherwood Forest, still haunted by the ghost of taking from the rich and giving to the poor. Soon enough the Establishment was properly rattled. Damaging stocking frames was made a capital crime.

King Ludd's soldiers were hardly lumpen primitives, or barbarians throwing themselves against the city walls of civilization, smashing what their inadequate brains failed to comprehend. On the contrary, their awareness of change was sharp and nuanced. Those at the head and heart of the uprising had watched and broadly tolerated the creep of technology, its floodwater

Felled in the battle between men and machines . . . mill owner William Horsfall is made a victim of Luddism

drowning others before reaching them. Accepting its advent as inevitable and irresistible, feeling their feet getting wet, they astutely demanded the right to have the machines serve them as well as the owners. In the main, the capitalists targeted with violence were those deemed to be using machines in 'a fraudulent and deceitful manner'. It was often time-served, skilled men who took up arms to protect their status, and if machines were to be used, then those with the skills should be employed at appropriate rates to supervise their operation.

It was not to be. Angered and properly alarmed by the burgeoning scale of it all, the government came down hard. On 24 March 1812 Luddites torched a factory in Westinghouse, Lancashire. In the aftermath, a trawl of likely suspects was hauled into court. Four were hanged and nine transported to Australia for seven years.

As well as craftsmen, the unskilled poor were Luddites too. Long years of war and its associated privations had left many with less than nothing to lose. Despair and desolation wept its last tears on the night of 9 June 1817 when around three hundred jobless labourers joined, or were effectively pressed into, a march on Nottingham. The leader was an unemployed stocking-maker named Jeremiah Brandreth (apparently an ancestor of Giles). Likely tricked by a government spy – in the pay of Home Secretary Lord Sidmouth and looking for a patsy to trigger an outrage that might justify a horror – Brandreth promised their hundreds would be joined by tens of thousands for a march on London. After the years of Luddite effort had come to nothing, the dream was to sweep, like a dark cloud, into London, and Whitehall, and there demand wholesale reform. A perfect storm made of the Corn Laws, which hiked the price of bread, tax on commodities to pay the war debt and the influx of demobilized soldiers in want of work meant lives barely liveable for millions were finally unbearable.

The plans were hatched in the village of Pentrich. The Pentrich Revolution was the name given to the ill-starred effort that saw Brandreth shoot and kill a farmer's servant when he refused to open the door to men in want of food and drink. All along the road to Nottingham it was the same – doors barred and no support. The promised thousands were a hoax. Even the heavens showed their disdain by opening upon the rebels – rain like stair rods fell on shouldered pitchforks advancing towards government muskets. On the outer limits of the city, waiting soldiers opened fire and charged. Men fled, desperate and drenched. Scores were arrested.

Tory Prime Minister Robert Jenkinson, Lord Liverpool, saw to it that forty-seven men were charged with high treason. Derbyshire landowners made up the jury that found them guilty later that year. Four men were sentenced to death and twenty-three transported to living deaths in Australia. Unlike the Tolpuddle Martyrs, none of the Pentrich men, luckless labourers all, ever returned. Conviction for high treason might have meant hanging, drawing and quartering, but the intervention of the Prince Regent saw Brandreth and the rest despatched by simple hanging. The job was done in front of a grimly silent crowd, watched over by more soldiers warned of likely trouble. The corpses were then laid one by one on the chopping board, their heads severed. All the families of the convicted men were evicted, their homes razed to the ground.

If skulduggery it had been, intrigue unto death by the spy and his masters, it had achieved its goal. The Pentrich Revolution was the last armed uprising in England. A poem attributed to Brandreth reads:

Every man his skill must try,
He must turn out and not deny.
No bloody soldier must he dread,
He must come out and fight for bread.
The time has come you plainly see
When government opposed must be.

I say we are haunted now by those Pentrich rebels and all the Luddites that went before them. They had put the fear of God Almighty into the authorities: at one point, some twelve thousand soldiers were deployed to cope with Luddite violence. A further tragic irony is that most of those were local militiamen, in large part unemployed weavers and other craftsmen-turned-gamekeepers in the face of their poacher fellows.

If no Luddite ghosts walk among the living now, then perhaps their unquiet spirit – their essence – has returned, a revenant out of an older, wiser world. In his 1829 essay 'Sign of the Times', Scottish historian and philosopher Thomas Carlyle warned of 'mighty change' driven by technology:

We war with rude Nature; and, by our resistless engines, come off always victorious, and loaded with spoils.

... When we can drain the Ocean into mill-ponds, and bottle-up the Force of Gravity, to be sold by retail, in gas jars; then may we hope to comprehend the infinitudes of man's soul under formulas of Profit and Loss; and rule over this too, as over a patent engine, by checks, and valves, and balances ...

Not the external and physical alone is now managed by machinery, but the internal and spiritual also . . . The same habit regulates not our modes of action alone, but our modes of thought and feeling. Men are grown mechanical in head and heart, as well as in hand. They have lost faith in individual endeavour, and in natural force, of any kind. Not for internal perfection, but for external combinations and arrangements, for institutions, constitutions, for Mechanism of one sort or another, do they hope and struggle . . .

The plight of the mass of the people – those characterized by our technocrats as 'useless eaters' – was felt then as now. Charlotte Brontë, eldest of the Reverend Patrick's surviving daughters, featured the difficulties of the Luddites in her 1849 novel, *Shirley*. Before fame as a poet beckoned, the already rich and powerful Lord Byron stood up in Parliament in 1812 to challenge the then Tory Prime Minister Spencer Perceval's bid to make 'frame breaking' a capital offence. It was his maiden speech in the Lords, and he used it to speak for the Luddites, declaring that this 'once honest and industrious body of the people' had become 'miserable men . . . driven by nothing but want. You may call the people a mob,' he said. 'But do not forget that a mob often speaks the sentiments of the people.' Whether or not those words were true, they fell on deaf ears. Perceval's Bill became law and many men took the drop as a result.

In his 1984 essay 'Is It OK to Be a Luddite?' American novelist Thomas Pynchon noted that by the bleak and rain-drenched summer of 1816, Byron was in the Villa Diodati, by Lake Geneva, in Switzerland, swapping ghost stories with the Shelleys (Percy Bysshe and Mary) and John William Polidori. Hardly limited to

Lake Geneva, the weather during what became known as 'the year without a summer' was the result of the eruption, the year before, of the Mount Tamboro volcano in Indonesia. Vast clouds of ash had risen into the upper atmosphere, blotting out the sun, lowering temperatures and occasioning near-perpetual rain. The Tamboro effect heaped yet more misery on those English folk already weakened by the enervating toll of war, rising prices and unemployment. Bad weather meant miserable harvests as well. Those literary types tucked up in their villa, however, were troubled only in so far as the appeal of sightseeing paled and they had to divert their attentions indoors.

'By that December, as it happened,' wrote Pynchon, 'Mary Shelley was working on Chapter Four of her novel "Frankenstein, or the Modern Prometheus".' Shelley, then just eighteen, famously began her most famous work after a vivid dream apparently inspired by one of those evenings of ghost storytelling:

Night waned upon this talk, and even the witching hour had gone by before we retired to rest. When I placed my head on my pillow I did not sleep, nor could I be said to think. My imagination, unbidden, possessed and guided me, gifting the successive images that arose in my mind with a vividness far beyond the usual bounds of reverie. I saw – with shut eyes, but acute mental vision – I saw the pale student of unhallowed arts kneeling beside the thing he had put together. I saw the hideous phantasm of a man stretched out, and then, on the working of some powerful engine, show signs of life, and stir with an uneasy, half-vital motion. Frightful must it be; for supremely frightful would be the effect of any human endeavour to mock the stupendous mechanism of the Creator of the world.

Hers was a vision, then, of a man, mankind, crossing a line into the domain of God, daring to make of life itself a commodity that might be both given and taken on a whim.

Pynchon added that the novel remained well worth reading

for all the reasons that we read novels, as well as for the much more limited question of its Luddite value: that is, for its attempt, through literary means which are nocturnal and deal in disguise, to *deny the machine . . .*

What is clear . . . despite the commonly depicted Bolt Through the Neck, is that neither the method nor the creature that results is mechanical.

In the same nineteenth-century world that witnessed the rise of King Ludd and his subjects, Shelley and Byron were awake, subconsciously or not, to an existential battle between mankind and mankind's creations, both real and imagined. Pynchon described how so-called Gothic fiction, like Polidori's *Vampyre*, another product of that villa and those ghost stories, and Horace Walpole's *The Castle of Otranto*, was summoned by a world struggling with change, when the Age of Reason was snuffing out old flames, like magic and religion: 'Blake's dark Satanic mills represented an old magic that, like Satan, had fallen from grace.' He added that Luddites and the Gothic novel 'Each in its way expressed the same unwillingness to give up elements of faith, however "irrational", to an emerging technopolitical order that might or might not know what it was doing.'

How many of us harbour Luddite fears now, and with perhaps better reason than those stocking-makers and weavers of the nineteenth century? The technology they railed against was ach-ingly simple, easily located and destroyed with basic tools. They failed nonetheless, and those of us fearing the worst of what is coming, courtesy of the latest digital technology and AI, will surely have no option but to step aside and make way for the future too. The challenge, the foe, the already present reality of it is as amorphous and hard to pin down as any wraith. Rather than stocking frames, it is the cloud that we face, the cloud of AI grown monstrous on an endless diet of our personal data.

Surely those Luddites walk invisible among us, nodding at what they have seen before and perhaps wondering at what they could never have imagined. We have transhumanism too, the promised merging of man and machine. Where Luddites could swing hammers and axes at their offending machines, we and our descendants face the prospect of unholy union with ours.

Carlyle saw it, whether he knew it or not: 'Men are grown mechanical in head and heart, as well as in hand. They have lost faith in individual endeavour . . .'

Pynchon's essay is nearly forty years old. In the face of the impact felt then of rapid advances in computer technology, he suggested that the 'ability to get the right data to those whom the data will do the most good' might mean 'we will cure cancer, save ourselves from nuclear extinction, grow food for everybody, detoxify the results of industrial greed gone berserk – realize all the wistful pipe dreams of our days.'

I read those words now against a 2023 backdrop of the threat of nuclear war over the plight of Ukraine, COP27 demands to reduce meat and dairy herds and the amount of land given over to farming, plans for vast expansion of mining for rare earth metals and minerals and its lethally toxic consequences, and wonder about the state of those pipe dreams. A perpetually ailing NHS and two years of Covid lockdowns sent cancer deaths through the roof here in Britain, leaving Pynchon's optimism altogether flailing.

In his 1967 book, *The Ghost in the Machine*, Arthur Koestler fretted over the thought of the modern human brain grown large from something older, baser, given to overwhelming emotions, like hatred and wrath. That older form haunts our modern selves, restless and unfulfilled. Ned Ludd's offspring, King Ludd's men, are the ghosts in our machines.

20

ERNEST SHACKLETON, *SOUTH* AND THE FOURTH MAN

'When I look back at those days I have no doubt that Providence
guided us, not only across those snowfields, but across the storm-
white sea that separated Elephant Island from our landing place
on South Georgia. I know that during that long and racking march
of thirty-six hours over the unnamed mountains and glaciers of
South Georgia it seemed to me often that we were four, not three.'

ERNEST SHACKLETON, *SOUTH*

'A deserted house I could live with
But one deserted by you?
Come clean, old haunter
Is that you not there?'

UNKNOWN

FOR THE FEW WEEKS BEFORE HE DIED, DAD STAYED IN HIS BED. BEFORE THAT, AND
for what seemed like the longest time, we knew he was dying.
He had cancer in more than one part of his body and had opted

against treatment. Anything the doctors might have done for him was no cure, just the possibility of prolonging his life. They could not say by how much, and what they had on offer was chemotherapy of a sort that would have ulcerated his mouth and the lining of his throat, making eating and drinking a painful chore. So he said, no, thank you. He was fine for ages, up and about, sitting in his chair to read or watch television, almost always with my mum and one of my sisters. They still went out for drives in the car, with Dad still doing the driving. For the duration of that strange limbo, he was his old self at all times. And then, at some point along the way, the way he walked alone even in the company of those who loved him dearly, he began to fail, and quickly. Eventually he took to his bed and, instead of having a nap and rejoining the fold, as he had been wont to do, stayed there.

When I visited, driving down from my home in Stirling, I would lie sometimes on the bed beside him. I told him things I wanted him to know. One day, almost on a whim, I said I wanted him to contact me after he died. I can only imagine how many other people, in that same situation, try to pull the same death-defying trick. Between us we agreed a word, one that mattered to him and me.

'If you can,' I said, 'use it, communicate that word, make me see it or hear it, and I'll know it's you.'

My mum is almost blind. Now in her ninetieth year she has had age-related macular degeneration for years. The macula is an area of the retina, a few millimetres in diameter, a little yellow patch that looks after the definition and clarity of the central part of what a person is looking at. Like everything else, the macula wears with age; sunlight takes its toll too, over the course of a long life. Anyone with macular degeneration, like my mum, steadily loses that central portion of whatever they are trying to see. Recognizing faces becomes all but impossible. Reading, watching a screen, you get the picture – unlike someone with macular degeneration, who mostly does not.

Months ago, she told me about a more recent phenomenon she had noticed – in fact, could not avoid. While looking at something – anything – her compromised view was further confused by a group of little boys, wearing shorts and coats. They were always together in a group, moving as though in a scrum. Mum would be trying to watch television, or reading, making the best of that peripheral area of vision, and always the boys were there, roaming around as one, and right in the way of whatever she was trying to watch. Unlike anything else in the centre of her vision, she could see them clearly.

While I was writing this, I asked her about it again and she said it was no longer just a scrum of little boys. Now she was seeing, in the background of every view, a red-brick house surrounded by a green lawn and a set of railings. Everywhere she looked – the television, out of the window, just eyes open and looking ahead while seated in her chair – there it was, the house. It wasn't even somewhere she recognized, just a random house. What remains of her vision, the rim around the outside, is leached of colour, faded like an old photo, but the bricks of the house, and the grass around it, are bright. Sometimes her view is obscured by shrubs in front of everything. All manner of distractions, out of nowhere.

'It drives me mad,' she said.

Her optician says hers is a common symptom of people suffering loss of vision: for want of as much stimulation as it used to get, the brain of someone with macular degeneration, for instance, conjures up images of its own as a diverting distraction, some-thing else to look at, if you will. I looked it up. Technically the effect is a product of what specialists call the 'Pathology of Bore-dom'. It doesn't just afflict those with sight loss.

Ernest Shackleton reminds me of my grandfather – in photo-graphs, I mean. That's not it exactly. My grandfather is hardly a doppelgänger for the legendary explorer. Truer to say Shackleton reminds me of my dad's side of the family, something about the wide face, the mouth, the set of the jaw. Maybe it is just my imagination, on account of my knowing both were born and raised in the same Victorian–Edwardian world. As I type this I am seeing, in my mind's eye, a photograph of Dad's dad, Robert Miller Oliver. It is the only one I have seen of him in his First World War uniform. In 1916 he joined the Ayrshire Yeomanry, which became part of the Royal Scots Fusiliers. In the picture his uniform looks uncomfortably stiff and new. Perhaps he had the picture taken on the first day he wore it. There's no one to tell me now, no one to ask. I have the gold wristwatch that was his retire-ment gift. The inscription on the back reads:

Presented to
R. M. OLIVER
In appreciation of
41 years
Loyal service with
Cooper & Co's Stores Ltd
1913-1954

Bound for war . . . Robert Miller Oliver, with his brothers, circa 1916

That always makes me stop to think as well – that after three years working in a grocer's shop in the south side of Glasgow, he was in the Great War. And when it was all over, he came back to the same job and kept at it until he was too old to work. Dad told me how, when Grandpa came home to his parents' house, his father took that uniform, filthy and stinking, and burned it in the back garden.

I was given his watch after he died, in the Erskine home for veterans, in Renfrewshire, in 1985. I never wear it – too afraid of it being damaged or lost. I used to wind it and watch it tick, but after a while I was too worried the day would come when I might overdo it, and the mainspring would snap. So it lies in a box of keepsakes close by the Crown and Eagle button from Dad's RAF uniform. Sometimes it bothers me that I leave those things in a drawer, all but forgotten. What is the point of a key fob without keys? What is the point of a stopped watch?

Grandpa's face in the photo amazes me – that's the only word, *amazes*. He was in his seventies when I was born, and ancient to me. Now that I am fifty-six, seventy doesn't even feel far off, far less old. But he was white-haired and stooped and reminded me of Private Godfrey in *Dad's Army*. If I ran a finger behind one of his ears (I forget which) I could feel, and almost see, beneath the thin skin, a serrated edge, an old bread knife mostly dulled. I asked Dad what it was and he said it was a fragment of a 'whizz-bang', a small explosive shell. Grandpa had been hit and wounded by an explosion but, for whatever reason, the doctor who patched him up left it in place. When I was at school, being taught about the war, the Somme, Passchendaele and the rest, I would think of

Grandpa and see him only as the old man I knew. There he was, marching along, white-haired and plump in his uniform, with tartan slippers on his feet on the road to Menin. That solitary photo of him turned up only recently, sent to my dad by some or other relative who had come across it among an all-but-forgotten pile of papers. Amazing: Grandpa was young once – fancy that!

The same year R. M. Oliver joined the Ayrshire Yeomanry and donned that uniform for the first time, Ernest Shackleton walked across the island of South Georgia, in the South Atlantic, on the last leg of a journey most epic. He and his crew – twenty-eight in total – had been gone a long time, having departed English shores in June 1914, bound for Antarctica. Their ship, *Endurance*, was gone altogether by then.

The plan had been to walk from one side of the continent to the other, via the South Pole, but the so-called Imperial Trans-Antarctic Expedition had been over before it could properly begin. Trapped in pack ice on their approach, *Endurance* had had her limits after all, the life crushed out of her before she sank on 21 November 1915. While she slipped into the dark, Shackleton told his men: 'So . . . now we'll go home.' They departed the ice in three little boats and, led by Shackleton and navigator Frank Worsley, completed a week-long crossing to the uninhabited rock that was Elephant Island. Since there was no hope of rescue from there, six men got back aboard the largest of the three boats (called *James Caird*, after the ill-fated expedition's principal sponsor), Shackleton, Worsley, Tom Crean, Tim McCarthy, Harry 'Chippy' McNeish and Jack Vincent. On 10 May, after seventeen days sailing eight hundred miles across the Southern Ocean in an open boat just twenty-two and a half feet long, guided only by Worsley's dead reckoning, they made it to South Georgia.

They had had only one chance: if Worsley had missed the target – just thirty miles long, a speck in the vastness – the currents and prevailing winds would have swept them into empty oblivion. So near and yet so far: having landed in King Haakon Bay, in the south, their only hope of finally finding the help upon which all lives still depended – their own and those of the twenty-two souls waiting beneath two upturned boats back on Elephant Island – was a whaling station at Stromness, on Grytviken Bay, in the north. McNeish and Vincent were spent. McCarthy stayed behind to care for them, while Shackleton, Worsley and Crean set out on foot to cross the glacier fields and mountains of South Georgia's then unknown and unmapped interior. After a non-stop, thirty-hour push across forty miles, climbing to heights of four thousand feet and picking their way over sheer drops and crevasses, they arrived at a whaling station manned by a crew of Norwegians. By late August, Shackleton had made it back to Elephant Island to collect the men he had left there.

Shackleton's example of leadership is immortal now, invoked again and again by those seeking to inspire in others an awareness of the potential of the human spirit. In the annals of heroics, there are stories that might equal the odyssey of the three – Shackleton, Worsley and Crean, who made it all the way – but none to surpass it. In *The Worst Journey in the World*, Apsley Cherry-Garrard, a veteran of Captain Robert Scott's 1910–13 Terra Nova Expedition, had noted Shackleton's grit: 'For a joint scientific and geographical piece of organization, give me

Scott; for a Winter journey, Wilson; for a dash to the Pole and nothing else, Amundsen: and if I am in the devil of a hole and want to get out of it, give me Shackleton every time.'

At Stromness, among the whalers, Shackleton and the others had told their story to some of the hardest men, who hunted the greatest beasts in the cruellest environment. When they were done, a veteran of forty years and more in the south stepped out of the tobacco-smoke haze in the cabin and told them it was an honour to have been in their company. All but lost for words, he gestured with the stem of his pipe towards the three: 'These,' he said, nodding, 'are men.'

In 2010 I was part of a seven-man crew that sailed a fifty-foot yacht (called *Pelagic*, which means 'of the open sea') from Port Stanley in the Falkland Islands to the South Orkney archipelago inside the Antarctic Reserve. We were making a television documentary about a Scottish pioneer called William Spiers Bruce, who mapped part of the continent of Antarctica between 1902 and 1904, long before Shackleton, Scott or anyone else had bothered to do so – and it was someone's bright idea that we should experience what the Southern Ocean had to offer. It took us more than a week to sail down and even longer to sail back – uphill, as it seemed to me – with a couple of weeks on the islands. I won't claim even to have sighted the hem of Shackleton's garment, far less touched it, but I have been on some of the same water, among grey waves as big as hills, sailing past icebergs grander in scale than whole city blocks. I saw the black backs of leviathans and the impossible silhouette of the wandering albatross.

I talk sometimes to some of those who were there, too, about how we knew a low-level, beneath-the-surface fear the whole

time we were at sea; about how the thought of all that cold darkness beneath the hull, the same that had swallowed *Endurance*,
weighed heavy on our minds; the anxiety of standing on the bow,
in the dark of night, peering ahead for fear of lumps of ice, called
growlers, that floated just beneath the surface, like our fear, and
that promised catastrophic damage for a yacht such as ours; that
the relief of arriving back in Port Stanley at the end was so intense
we lost track of time for an entire day. And yet we all say the same
thing: I want to go back. I've been lucky enough to have taken all
sorts of trips, to all sorts of places hot, cold and in between. But
the one I see most clearly in my mind's eye, the one to which I am
transported every time I see a yacht in full sail or catch a glimpse
on TV of wind-ripped grey water and a boat's bow rising and falling, the one that haunts me, is my round-trip to the end of the
world, a place like no other, no other at all.

What haunts me most of all, I think, is the lingering essence of
the feeling of being in a place beyond the reach of help. They say
some solo round-the-world yachtsmen and -women don't bother
with life jackets in the South Atlantic: if they fall overboard there,
where no help is coming, buoyancy aids just prolong the agony of
exposure. Maybe that's a story to scare newbies, but I close my
eyes and remember the emptiness of that water.

On his return from Antarctica, Shackleton wrote an account of
it all, called *South*. When it came to that slog across the wastes of
South Georgia with his two compatriots, he described an encounter or, rather, the presence of another:

> When I look back at those days, I have no doubt that Providence
> guided us, not only across those snowfields, but across the

storm-white sea that separated Elephant Island from our landing place on South Georgia. I know that during that long and racking march of thirty-six hours over the unnamed mountains and glaciers of South Georgia it seemed to me often that we were four, not three.

I said nothing to my companions on the point, but afterwards Worsley said to me, 'Boss, I had a curious feeling on the march that there was another person with us.' Crean confessed to the same idea. One feels 'the dearth of human words, the roughness of mortal speech' in trying to describe things intangible, but a record of our journeys would be incomplete without a reference to a subject very near to our hearts.

The 'dearth of words' is from Keats's 'Endymion' – 'Some shape of beauty moves away the pall/From our dark spirits . . .' Explorers reach with every fibre for the unknown, the as yet unseen. Poets reach for the intangible. *South* was published in 1919 and three years later T. S. Eliot echoed the thought of the fourth man in *The Waste Land*: 'Who is the third who walks always beside you?'

I think about my dad and his vision of a guardian angel standing over my mum that night long ago.

It is from Eliot, therefore, and not from Shackleton, that the phenomenon takes its name. A third man or a fourth. The phenomenon is experienced just as often by people otherwise alone. In his 2008 book, *The Third Man Factor: Surviving the Impossible*, John Geiger relates numerous accounts of people *in extremis*, facing death, yet feeling an invisible presence that offers hope and sometimes help. On 11 September 2001 Canadian Ron DiFrancesco was at his work at Euro Brokers on the eighty-fourth floor of the South Tower of the World Trade Center. He and his colleagues had already heard an explosion from the North Tower. Phones were ringing. There was talk of a light plane having lost its way, a tragic accident. They could see fire and smoke, people leaping to their deaths to escape the flames. Some of his colleagues began leaving the South Tower, but at first Ron stayed put. A message came over the building intercom, saying the building was secure, everyone could return to their offices. His phone rang, a friend from Canada. 'Get the hell out,' he said.

Having heeded that instruction, Ron was at the elevators waiting to go down when, as we are told, the second plane, a Boeing 767, travelling at nearly six hundred miles an hour, hit the building. The one that crashed into the North Tower had been in level flight. The second hit the South Tower at an angle, so that fire broke out both above and below the eighty-fourth floor. The whole building swayed. Ron thought it was going to topple, but it shuddered back into place. He and others made for the stairs. Smoke billowed. Grabbing a piece of drywall from among the debris, he used it as a shield while he tried to descend. The heat was too much, and he stopped and lay down to breathe clearer air. Everyone else turned to go back up, seeking safety above the flames. It was at that moment, lying on the floor, that Ron felt lifted up. He heard a voice too, telling him to keep going.

While his colleagues turned back, Ron headed alone into the smoke and flames. The voice gave him faith and he followed where he was led. There was an intense sensation, he said, not just a voice but a physical presence. Now there was a second voice from some floors below. He was on the seventy-sixth by then and it was as he realized he was within reach of help that he felt the invisible other leave his side. Ron reached the source of the new voice: a fireman, one of three, and told them he couldn't breathe. Still on their way upwards, one checked him briefly, assessed him as OK, and told him to keep going down. On the floors below, the sprinklers were on, making life easier. Finally reaching the ground floor he was intercepted by more firefighters before he could get out on to the street. It was too dangerous, they said. He looked and saw bodies and debris raining down on to the ground outside. Directed to the basement he was joined by a second man, another survivor. As they made for an exit there was an explosion and Ron turned to see a fireball. Both men ran.

Ron regained consciousness in hospital. As far as anyone can tell, he was the last to escape the eighty-fourth floor, one of the last to leave the building alive.

He insisted afterwards that it had been divine intervention. In the years since he has struggled with his mental health. In 2021 he told the *Toronto City News* he has survivor's guilt: 'I was in the stairwell with a bunch of them and left them there,' he said. 'I followed a voice and got myself out, but I know if I had gone back to get them, I wouldn't be here today.'

Geiger's book records story after story: mountain climbers, sailors, survivors of catastrophes natural and man-made, all have reported help of the sort described by Shackleton and DiFrancesco. Psychologists suggest the phenomenon is the product of a coping mechanism, summoned unconsciously from within rather than arriving from without, a manifestation of nothing less than the human spirit.

As well as my dad's story about my mum's guardian angel, I think of her sister, Elma, and Sonny, the invisible friend who was there in the years after her infant brother died and her family grieved his loss. Who can say now what walked or did not walk beside Shackleton, Worsley and Crean? Who can say what lifted DiFrancesco in a stairwell of the South Tower and led him to safety? Ghost? Guardian angel? Figment of the imagination?

It is, anyway, an old story. In the New Testament, in Luke's gospel, two Apostles are walking along a road to Jerusalem in the aftermath of the crucifixion. They are joined by another:

> And it came to pass, that, while they communed together and reasoned, Jesus himself drew near, and went with them. But their eyes were holden that they should not know him.

In the worst of times, in grief and in mortal danger, we might reach out for help, the name and source of which we do not know. Eliot wrote *The Waste Land* in the aftermath of the First World War. In part it reads like a lament for all that had been lost, perverted or destroyed in that very worst of times. It is a poem haunted by the irretrievable past, the past that has been shattered into fragments – a reminder that we might inadvertently summon and be accompanied for a time only by sadness.

Throughout every ghost story the ghost is yearning for something out of reach and out of sight. That yearning comes from us, is carried with us from haunted place to haunted place. The third man, or the fourth, whichever, might be the epitome of what it means to be haunted, to encounter revenants of those we can no longer know, no matter how much we might want to, when all we really have is inside us, made of everything we have seen and all those we have loved.

21

COFFIN ROADS

'Now it is the time of night
That the graves, all gaping wide,
Every one lets forth his sprite,
In the church-way paths to glide.'

WILLIAM SHAKESPEARE, *A MIDSUMMER NIGHT'S DREAM*

IN 2019 A FORMER DIRECTOR OF PUBLIC HEALTH FOR THE NORTH-WEST OF ENGLAND suggested that that nation's dead might be laid to rest alongside motorways, railway lines and public footpaths. Professor John Ashton CBE said the time had come to find new, more environmentally friendly ways to dispose of the mortal remains of the half-million or so who die in England and Wales every year and proposed 'green burial corridors' planted with trees. One for each of the dead: half a million trees a year. Ashton's vision of linear cemeteries strung out along the nation's transport routes put a new spin on the ancient tradition of coffin roads.

Also known as bier roads, corpse roads, church ways and lych ways, coffin roads were spun like a spider's web, fine strands linking often remote communities to their nearest hallowed ground. In old days the Church and the Kirk held sway over the dead as well the living. In every parish there was a mother church surrounded by its consecrated ground. Such clay was jealously guarded, not least because burial rites meant money for church coffers but also on account of the perceived need to gather up the bodies of the dead that they might be counted and kept safe until resurrection. The good shepherd must mind his flock, dead or alive. Populations grew over time and individuals and families settled on land further and further from the centre. But while satellite churches might be built for worship, only the mother church held ground blessed and fit for cradling the loved bodies until such times as Jesus would return to raise them up.

The principle was straightforward enough, but in practice the carrying of a coffin over hill and dale was a physical and an emotional challenge. If the death occurred in winter – when the weather was cruel, snow drifting, storms baying – the body might have to be kept, naturally chilled, in an outhouse or ice house, until such time as transport was possible. Then as now, folk were superstitious, especially about the dead, so were disinclined to see coffins pass close by, especially not over their land. Coffin roads were therefore laid over remote terrain, out of sight, meaning they took less than direct routes from A to B. Such ways, over wild terrain, were usually unfit for wagons and carts so that coffins were carried instead by relays of strong men drawn from scattered communities. The whole effort might take days, the pall-bearers provided with food and bedding transported on pack animals. According to John Hillaby, writing in a December 1977 edition of

New Scientist: 'The belief was that coffins sterilized the land because the dead were forced to walk that way until their souls were purged.' To prevent as much, flat stones were laid alongside the corpse roads as resting places for the coffin, lest it touch the ground. Given the dread associated with such ways, they were seldom used by the living for any other purpose and were regarded as best avoided. And that which exists only at the edges of vision, and of attention, is often the source of strangeness.

Like the Lych Way from the valley of the Dart (by Wistman's Wood, the Wild Hunt and the Hounds of Annwn, to Lydford): a famous coffin road winds from lonely Swindale Head to the glacial valley of Mardale, in the Lake District, Cumbria. Fell-walker Alfred Wainwright, he of the seven-volume *Pictorial Guide to the Lakeland Fells*, vented much spleen when Mardale was dammed and flooded in the mid-1930s by the Manchester Corporation to make Haweswater Reservoir. England is haunted by the ghosts of many villages lost to lakes man-made. Beneath the water of Ladybower Reservoir in the Peak District, where phantom Lancaster bombers fly silent for those who can see them, lie the shades of Ashopton and Derwent.

The hamlet of Mardale Green was drowned by Haweswater. Along with some farms and the Dun Bull Hotel, there was once a church. Its mother was at Shap, in the county of Westmorland, where an abbey of Premonstratensian canons held sway until the dissolution by Henry VIII. Mardale Green was granted first an oratory, and then a church, the Holy Trinity, big enough for seventy-five worshippers. For the longest time the dead of Mardale Green had to be carried via the coffin road to Swindale Head, hard miles, then onwards for more to Shap. Sometime in the eighteenth century Mardale Holy Trinity was granted permission to bury her own dead and did so for two centuries.

With the water slowly rising, Mardale Green was abandoned, most of its buildings blown to smithereens by men of the Royal Engineers using it for practice. The church was left intact for a while, a final service held on 18 August 1935. Records show that seventy-two people gathered inside to hear the Bishop of Carlisle offer his blessings upon the place. A thousand more, drawn from all over Westmorland, listened to the proceedings outside, broadcast on loudspeakers. The church was taken apart then, stone by stone, some reused for the structures of the reservoir. All ninety-two graves in the churchyard were reopened, the remains within removed to Shap. After all those years, and all those funerals, the mother church had claimed them after all.

Coffin roads are everywhere, or were, some lost beneath modern roads and public rights of way. Like long grey hairs shed and all but lost in the undergrowth, proof of lives lived and also of inevitable deaths. In our world of last breaths exhaled in sterile hospitals, of discreet undertakers carrying off corpses for washing and dressing by strangers, of no-strings cremation services advertised on daytime television, we are mostly spared the immanence of death. Those old ways, though – lych ways, bier roads, corpse roads – are muffled reminders of a time when we were carried not by strangers but by our families, from cradle to grave. Helpless on the way in and helpless on the way out but known and always held.

Also in Cumbria there is a coffin road from Rydal to the nearest consecrated ground in Grasmere. Rydal Mount was home to William Wordsworth from 1813 until 1850 and tradition has it that his spirit returned there after death, appearing in the corner of his sister Dorothy's bedroom. A person does not have to be superstitious to sense an atmosphere suspended like mist from the branches that overhang that winding way. Loved by those who knew them, each of the dead became in time just part of the fellowship of the ancestors. Folklore had it that restless spirits took the shortest path from grave to home so coffin roads were deliberately winding to throw them off. Coffins were carried so that the corpse's feet went first along the way, to make them less inclined to walk back towards home. Spirits were disinclined to cross running water, so coffin roads often made a point of fording shallows en route to the cemetery. They had a loose grasp of direction, too, apt to be confused by junctions, so coffin roads often made a point of passing over a crossroads. Since all those efforts might confine a wandering spirit to the road, those ways were feared, avoided, and tales of hauntings were inevitable.

The Swaledale Corpse Way in Yorkshire links the village of Keld, at the top of the valley, with Grinton and the consecrated ground there of St Andrew's Church, sixteen miles lower down. When the church of St Mary the Virgin was built at Muker, in 1580, closer to Keld, the old way fell into disuse. Still to be seen, though, are the numerous flat 'coffin stones' that were laid alongside the way as safe resting spots. Anyone pausing too long at the coffin stone by the River Swale is apt to glimpse a black dog, headless, as it happens, and regarded as a portent of a death in the family. To keep the weight down along the way, wicker was used for the coffins, rather than wood. Since wool was important to the local economy, laws were passed in the latter part of the

The way of the dead . . . Ivelet Bridge over the River Swale on the Corpse Way

seventeenth century to bar the use of new-fangled fabrics for winding sheets. At Grinton Church a stone slab records the five-pounds fine paid by one Adam Barker for burying his daughter, Ann, wound in linen.

Along the Swaledale and other coffin roads there is talk of coffin stones that talk to passers-by. Elsewhere there are reports of 'corpse-candles' – amorphous lights coloured yellow, or sometimes blue, emitted no doubt by the spirits of the dead as they wind their way back and forth along the paths taken by their coffins. In Wales the lights are *canhwyllau cyrff* and, rather than manifestations of the dead, they are regarded as portents of death. In *South Wales Ghost Stories*, Richard Holland quotes the nineteenth-century historian James Motley's 1848 encounter. Of the lights he said they seemed to be of electrical origin, when the ears of the traveller's horse, the extremity of his whip, his spurs or any other projecting points appear tipped with pencils of light . . . the toes of the rider's boots, and even the tufts of hair at the fetlocks of his horse, appeared to burn with a steady blue light, on the hand extended, every finger immediately became tipped with fire.

Even by the middle of the nineteenth century therefore, rational minds were brought to bear upon phenomena once regarded as only supernatural. But still the traditions persisted in the recollections of the living. On the website of Amgueddfa Cymru (Museum Wales) there is a recording of one Mary Thomas, of Ffair-rhos, in Ceridigion, west Wales, describing her grandfather's experience of corpse candles:

'He'd had many experiences of the corpse-candle. My grandmother died when my mother was eight years old . . . She died of the *diacad* [tuberculosis], as they called it in those days . . . there was no cure. And the night before she died he was by her bedside, and he saw a little lighted candle on the bed, and he saw it going out of the house. And then his wife died. And he saw his wife's corpse-candle going out of the house. And she saw it too. She said: "Do you see that light going out through the door, Tomos?" Both she and he saw the light, and she died the next day . . . Oh, they certainly saw it.'

As is the case with every mention of a ghostly or otherworldly encounter, it is easy to dismiss talk of corpse-candles, headless dogs and talking stones as flights of fancy. But perhaps it is important at least to consider such experiences in the context of their place in the world, if not their time. As well as part of phys-ical landscapes, coffin roads and corpse-lights emanate from worlds shaped not just by climate and geology but also by culture and paideia. That which was real to the people who experienced it (or thought they did) is worthy of consideration for the evi-dence it provides of beliefs held, lives lived. Still within reach of the living are worlds in which belief and faith in the invisible are as real to people as trees and streams, as sound as the paths they walked. Like the water in which a fish swims, oblivious to the water's existence, so people are in and of the culture in which they live their everyday.

The same thinking that prompted the suggestion of burying bodies alongside motorways and railway lines makes plain the polluting nature of modern burial. Toxic chemicals in the embalming process and the materials used in the traditional coffin – resins, glues and plastics – are all bad news for the ground into which they are placed, making modern cemeteries sterile at best for centuries to come.

Consider, too, the valuable resources involved – used once, then put out of sight and out of use for ever – foreign hardwoods, bronze, copper and other metals in coffin details. In the West we are told, by voices louder and louder, that all such behaviour is wasteful at best and heinous at worst. But in ages past, the dead were laid down with wondrous things of great value: swords of bronze and iron, jewellery of gold and precious stones. To do anything less might have been considered disrespectful to say the least. Now we throw such wealth away without a thought. Just because we do not share the same approach now, does not make it right or wise merely to dismiss the lived realities of those to whom it meant everything. And so the world of the coffin roads, and belief in corpse-candles and homesick spirits, was real to its inhabitants.

Scotland is marbled with its own tracery of ways of the dead. The tiny island of Iona has its Sraid nam Marbh (remember?) from Martyrs' Bay to the burial ground of Reilig Òdhrain, by the abbey. Years ago, I visited the abandoned island of Scarp, to the west of Harris, off Scotland's west coast. It is a scrap of a thing, three miles across, and separated from its bigger neighbour by half a mile of Atlantic Ocean called the Sound of Scarp. Some families eked a living there until 1971 when the last were evacuated to Harris, life having got too hard. Scattered houses are reclaimed by the elements and time, although a few are repurposed as holiday homes. Local legend has it that the Moody Blues were on retreat in one when they came up with 'Nights In White Satin'.

On a knoll of high ground is a cemetery. There are some gravestones – including two of those distinctive Portland slabs preferred by the Commonwealth War Graves Commission. Donald MacLennan – Domhnall Ban, Fair Donald, as he was known, for his blond hair, and to differentiate him from others of the same name – was forty-four when he died on 15 November 1918. Nearby, as if to make the point about names, is Donald John MacLennan, apparently no direct relation. He died on 18 March 1917, aged thirty-two. Fair Donald was home on leave when he died of a stomach illness. D. J. MacLennan made it back somehow, even though British Army policy was to bury men where they fell. That they both lie in Scarp burial ground says much about the gravitational pull of home soil.

A Scottish (or, some say, Celtic) tradition makes a virtue of burial in the west, the westernmost point, since it is closest to the setting sun. Most coffin roads start in the east and finish in the west. Among the ancient Irish, the fabled Tír na nÓg, the land of the forever young, was somewhere in the far west too, and throughout the Middle Ages it was often to the islands of the west that the great and the good were carried for their eternal rest. It was practical as well: the geomorphology of the islands of the west meant the soil in the east was thin over rock, making it a necessity to carry the dead into the west coast and the machair, where might be found the fertile sandy soil between the land and the sea. Scotland's west is frayed and fretted with fiddly inlets, islands, isles and islets – a Lordship of Isles right enough – so that boats always made more sense than journeys over land.

Those returning to an island like Scarp for burial must come by sea, of course, but it is a coffin road they follow, just the same. There are still a few folk of Scarp out in the wider world, still breathing. Perhaps some or all will go home in death to lie beneath the long grass. Apart from the conventional gravestones, most of the resting places in that loneliest of cradlings are marked with half-buried beach pebbles, pale like skulls. Likely it was that way for the longest time. If the dead of Scarp are remembered, it is by a few still living. It is a haunted, haunting place and if ghosts roam anywhere, pining for those who ever loved them, it is there in the west of the west of the west.

The Stoneymollan coffin road winds from Balloch, on Loch Lomond, to the burial ground of St Mahew's Chapel, at Kirkton by Cardross on the Firth of Clyde. St Mahew's was restored in the 1950s but has a history that wanders all the way back to 1467.

Details are hard to come by, but the site may originally have been made sacred by a sixth-century saint named Mochta (not Matthew, for anyone wondering). Another coffin road nearby is that which was used by the clan MacFarlane, of Arrochar. Since their mother church was at Luss, the chiefs (and those who warranted the effort) were carried along the String Road, over the pass to Invergroin, in Glen Douglas, then onwards to the churchyard by Loch Lomond. Before departing Upper Glen Luss, for the final stretch, it was the clansmen's custom to leave their weapons behind at Cnoc an Airm to stand in the shadow of their maker unarmed.

Another way of the dead linked Kilcreggan, via Portkil, to Rosneath churchyard, and there are more besides, lost beneath the green. That those tracks remain, in memory if not in fact, linking places unfamiliar and unknown to all but a relative handful today, is testament to how long they mattered. Hilaire Belloc was surely right when he identified the road – The Road, even those roads taken by our dead – as the oldest and most venerable of our ancestors' creations, 'the greatest and most original of the spells which we inherit from the earliest pioneers of our race … the most imperative and the first of our necessities . . . before we were quite men, we knew it . . .'

While it was upon flat stones that coffins were laid elsewhere, among the Scots it was often cairns that served to keep the dead from settling into earth and so spoiling the lands through which they were carried. Mourners following might join the pall-bearers for a drink and a bite and, before moving onwards, add more stones to the ever-growing pile.

Sith do d'anam, is Clach air do Charn: Peace to your soul, and a stone to your cairn.

As is true elsewhere, there are innumerable tales of hauntings of the Scottish coffin roads. Just as the Brahan Seer saw, with his second sight, the end of the Seaforth line, the butcher's yard of Culloden Moor, so others with his gift witnessed ghostly processions on those ways of the dead. Always they were regarded as premonitions of funerals to come.

In all manner of ways, the coffin roads give rise to strangeness. On Harris, a coffin road provided passage on foot from Loch Airigh and the Bays of Harris on the east coast to the deep and sleepy soil of the machair of the west – and therefore the cemeteries at Borve, Luskentyre and Scarista. On one occasion, they say, a funeral party was carrying its burden across the inland moor, past lochans as shining dark as Japanese lacquer, towards Luskentyre, when a noise was heard from within the coffin. Opening it they found its female occupant alive and well, so she was carried, still in her coffin, all the way home to a family soon convinced of a miracle.

Life and death travel side by side, or life is the way and death the destination. Consciously or unconsciously, coffin roads were part of keeping death close, of confronting its constant presence. We were familiar with death, acquaintances not strangers, and not so long ago. Once upon a time it was part of the duty of a newlywed bride to prepare two winding sheets, one for her husband and one for herself. The body of a dead loved one was kept at home until the grave was ready. Washed and dressed, hair brushed, he or she would be laid in bed, or else in an open casket in a room of the house, for mourners to visit. The final journey along the coffin road was an arduous distraction from grief to come, necessary effort leaving those grieving no option but to deal with the land of the living. There was communion between the living, the dead and the landscape from which all had come and to which each would return. Ghosts are lesions on the flesh of the life. We need ghosts just as we need life, just as we need death.

22

The Skye Ferry and the Wee Black Car

'Sure by Tummel and Loch Rannoch and Lochaber I will go
By heather tracks wi' heaven in their wiles.
If it's thinkin' in your inner heart the braggart's in my step
You've never smelled the tangle o' the Isles.
Oh the far Cuillins are puttin' love on me
As step I wi' my cromach to the Isles.'

KENNETH MACLEOD, 'THE ROAD TO THE ISLES'

'A man who is not afraid of the sea will soon be drowned, he said, for he will be going out on a day he shouldn't. But we do be afraid of the sea, and we do only be drowned now and again.'

JOHN MILLINGTON SYNGE, THE ARAN ISLANDS

THE HEBRIDES OF SCOTLAND ARE EUROPE'S LARGEST ARCHIPELAGO. THE WHOLE – A careless scattering, like the fragments of a dropped and shattered plate – is split into the Inner, nearest the mainland, and the

Outer, further flung into the Atlantic. Skye is the largest of the Inner Hebrides and so close to the mainland (just three hundred yards offshore) as to be readily mistaken for a peninsula of same.

From the landlubber, car-centred perspective, all islands appear cut off, awkward places. In days gone by, however – ancient days when swathes of the inland were dense woodland or trackless morass, mountains steep, valleys dark and all but impassable on foot or by horse – the best of all ways to travel was by boat around the edge. When interiors were impenetrable and forbidding, coastlines and nearby islands were downright welcoming. Proficient in boats of necessity, islanders had the best of it.

In the first century BC Diodorus of Sicily wrote, in his *Historical Library*, about an island in the north 'beyond the land of the Celts' that was home to those he had learned to know as 'Hyperboreans'. Visited by Greeks (according to Diodorus), the island was the site of 'both a sacred precinct of Apollo and a notable temple'. Innumerable writers have since speculated that the temple referred to was Callanish, a Neolithic setting of standing stones overlooking the waters of Loch Roag, on the west coast of the Isle of Lewis in the Outer Hebrides. Diodorus never travelled to see any of it with his own eyes. Rather, he was working from stories and texts that long predated his own time, so that those Hyperboreans were known to his world even earlier. If Diodorus and his predecessors knew of Avebury or Stonehenge – much further south and closer to the European mainland – he did not bother to write about them. Perhaps word of Callanish (or whatever it was named then) had reached Sicily earlier because it was on the seaways of the ancient world and well known when landlocked Stonehenge and other inland sites were still out of the way and remote. The Hebrides were long on the map of the world, imbued with essences of olden days, essences that linger and affect the living.

I am old enough to remember the Isle of Skye as a place reached not by bridge but by boat. On the first trip Dad and I took together into the Highlands, past the battlefield of Culloden and the grave marker of my Camerons, I was eleven years old or thereabouts. We had hurtled north from home in Dumfries in the yellow VW Beetle I remember so clearly, number plate and all, and I do mean hurtled. Since he spent a good part of his working life as a travelling salesman it was as well that Dad enjoyed driving. For years he sold plant hire – he rented diggers, bulldozers, cranes and the rest of the monsters needed for big engineering projects. Not that he ever owned such kit: he worked for those who did. For a while he was employed by a mercurial man named Paddy Hearn, whose moods, Dad insisted, were subject to the phases of the moon – calm for most of the month, mad as a cut snake for the rest. Dad spent some years working out of Fort William, in the Highlands, which meant him spending the working week there, away from home in a hotel called the Nevis Bank that we later stayed in together.

During that time, he acquired a precise knowledge of the roads back and forth to our house in the south-west. When it came to our trips into the Highlands, and there were several in my early teens, he could drive them like a local. Second-hand Beetle or not he would, when the feeling moved him, entertain us both by exploiting his foreknowledge of the twists and turns to pass car after car. I don't suppose my mum would have approved but we were a canary-yellow blur. I like to think our passing lines of more leisurely drivers was memorable to more than just me.

In those far-off days of the early 1980s the principal ferryboat to Skye left the mainland at the Kyle of Lochalsh for the roughly third-of-a-mile crossing to the village of Kyleakin on the island's eastern shore. I fell for Skye then and there on that first visit. It can be a hard place for some, hard to love: the Cuillin, so often cloud-wreathed, seems to conjure more rainfall than a person might usefully want, and most of my memories of time spent on that island are, truthfully, as damp as an otter's pocket. That said, even when the looming clouds are low, puncturing their bellies on jagged summits, it is still a place of moody beauty. Skye's is an ancient face, as lovely in old age as in her long-forgotten youth. And when the sun shines, hers is a fresh-faced Heaven upon this Earth, bright as a newborn.

The road from Kyleakin passes through Broadford, then onwards to Portree, the island capital. From the 1930s onwards that same road I travelled with Dad all those years ago was haunted, from time to time, by a little black car.

I am haunted by wondering if I have spent enough time with my children, seeded enough memories of me inside their heads. Two are adults now, or near enough, the third closing fast. I don't need the second sight of the Brahan Seer to see ahead an emptier, quieter house eroded by their absence. What places might they visit again and see me there in their minds' eyes?

*

Roads in the Highlands, and on Skye, can be demanding. Some seem casually slung, like lengths of frayed rope, landing where they will across the landscape. Often they are single-track, with the pregnant pauses of passing places made for waiting, to let another go by before proceeding. All in all, the effect is to have a person pay attention to the road ahead. Before and after the Second World War, locals and visitors alike began reporting sightings of a little black car approaching on the road ahead, sometimes travelling fast. Those with a keen eye for make and model often swore it was a 1934 Austin. Encounter after encounter unfolded the same way: the wee black car would be coming on, then perhaps a bend in the road concealed it from view or else it was swallowed by a stretch of dead ground . . . whatever. As often as not it seemed to be driven hard – hard enough to have the other driver pull in to the side or a passing place, ready to give way. And then . . . and then nothing. The wee black car never arrived alongside, never passed. It was just gone, vanished into thin air. Only on one occasion did a driver see it come by, and when it passed, a glance inside to see who was driving revealed it was . . . without driver or passengers.

It went on so long, this time of sightings of the wee black Austin, decade after decade, it became a game of local children to look out for it wherever they went. Visitors checking into hotels and guesthouses might mention an encounter, only to be met with an understanding nod, and an account of a shared experience. The wee black Austin was sighted on various stretches of the road, but most often on the approach to the ferry at Kyleakin. For a long time it was a tale for the fireside, a mystifying legend.

Sometime in the 1960s, there came a day when it all turned.

The usual handful of cars and foot passengers began boarding for the five-minute crossing to the Kyle of Lochalsh. A little black car was in the line-up, a local Presbyterian minister at the wheel. It was not his vehicle but belonged instead to one of the two young women he was travelling with that day. There was a young child in the car as well. The woman driver was nervous of the necessary manoeuvring and had asked her male passenger to take over. No one can be sure why events unfolded as they did. Perhaps the man, on the large side apparently, found the seating cramped and his feet too close to the pedals. In any event, the car shot forward, travelled the length of the deck, and collided with the upraised ramp at the other end. It dropped into the water and the car followed. The water was relatively shallow where it landed, but the car was still submerged up to its roof. A quick-witted crewman grabbed an axe and leaped from the deck on to the roof and tried desperately to break the windows. He smashed one and the minister, big or not, managed to escape. Despite further efforts, both women and the child were drowned. The car was black, right enough, and in the aftermath, they say the minister was driven quite mad with grief and guilt.

The second sight of the Gaelic Highlands, the burden of the seers, offers glimpses of the future to those blessed or cursed with it. Many are those who suggest the car that haunted the road to Kyleakin for all those years was somehow a shared second sight, a forewarning to many of a grim day to come. Since the tragedy on the ferry, the wee black car has never been seen again.

How to understand the legend of the wee black Austin on the road to Kyleakin? A phantom years before the tragedy that claimed three lives and destroyed a fourth. Was it a warning? And, if so, how on earth might it have been meaningfully interpreted by the hundreds who saw it over the decades? The facts remain nonetheless: uncounted numbers of people claimed to see it approach and disappear. And then the grim horror in one small black car at the end of the road where its image was most often seen. In some ways it is the story that, more than any other, gives me pause.

There is something about islands, anyway, and sailings into the west of the west. Dead kings to Iona; Arthur to Avalon; hobbits Bilbo and Frodo to the Undying Lands. Tír na nÓg, the Land of the Young. Places for going to and not returning, or not completely. Something brought, something taken.

Also in the Inner Hebrides is Jura, likely named for the red deer that roam – as in the Norse Dyrøy. It is sparsely populated, made in the main of bare rock and blanket bog. In May 1946 journalist and author George Orwell took up residence in a remote farmhouse there called Barnhill. It was on the point of slipping finally into ruin, in need of life, so its owner, Robin Fletcher, laird of Jura, was happy to see it rented – even if it had to be to a London leftie literary sort. Orwell arrived with his sister Avril, and was soon joined by his son, Richard, and their housekeeper, Susan Watson. His wife, Eileen, had died during a hysterectomy the year before. He came with personal sadness, then, as well as all the shared shell-shocked misery of that first year after the war.

As Jura is remote in regard to Scotland, so Barnhill is remote in regard to Jura: two ferries (a first to Islay and a second to Jura) then twenty miles by car and a four-mile walk along a rough track. All supplies – food and fuel and aught else besides – had to be carried in. For all that it might sound grim, Orwell loved it. For a start it fulfilled his dream of a Hebridean island on which to write the novel that would be *Nineteen Eighty-Four*. It also meant freedom, far from the madding crowd – he called it 'an extremely un-get-atable place' – a chance to grow food, hunt game, fish, and so live what he had imagined of other men's lives.

Barnhill is in the north of Jura, and a few miles off the western coast is the whirlpool called Corryvreckan, a Gaelic word meaning 'the cauldron of the plaid'. Celtic myth would have it that it is there in the roiling waters of the whirlpool that the Cailleach Bheur, the Hag of Winter, washes her clothes. Orwell and

The whirlpool of Corryvreckan . . . the cauldron of the plaid, where the Hag of Winter washes her clothes

Richard were in a boat in the Gulf of Corryvreckan one day when it was caught in the fringes of the maelstrom and pulled out from under them. By luck the pair escaped, struggling finally to a remote cove from which they were rescued by a fisherman. I think of the father and son drowned off Sandwood Bay, in Cape Wrath, and found lashed together on the sand: death is always closer than a person might allow.

Among it all he set to work on his final novel, his swan song. Before he left London, friends had noted he was gaunter than usual, frailer. For all that he was middle class and lived a softer life than the proletariat he had written about, life had taken its own toll. In the previous decade he had fought for the Communists in the Spanish Civil War and been shot in the throat for his trouble. He wrote up his experiences as *Homage to Catalonia*, published in 1938. Like all who lived through the privations of the Second World War, he was worn thinner than he ought to have been; and then the grief of losing his wife. By the time he was on Jura he was ill, dying of tuberculosis, though he did not yet know it. Sometimes he was well enough to write at the table in the living room of Barnhill, or he took to his bed and wrote there.

He was dying, but in those months on Jura he saw the future as clearly as any Highland seer, so he dreamed and typed a dreary dream of hopeless citizens in a hopeless world of bad food, bad clothes and perpetual war stoked by governments bent on totalitarian control; a world where war consumes the otherwise pointless products of the soul-sapping factories in which citizens spend their years; where there is no privacy, the citizens are watched always by Big Brother; a world where children spy on their parents; where power laid bare of all pretence lies in pain inflicted upon others; where the language of Newspeak denies to all the words that might give expression to free thought.

George Orwell (a *nom de plume*: Eric Arthur Blair is the name on his gravestone) saw the future, this future looming now, while he typed and coughed red blood in a white-painted house on an island at the edge of the west. He came and went from Jura, returning to London for the winter of 1946/7 and then back to the island in April. The first draft was completed by November, after which he sought treatment for his ills in Hairmyres Hospital in East Kilbride, near Glasgow. It was there that he received his TB diagnosis and ought to have dealt with it. Instead, he returned once more to the island of his dream. There he typed up his final draft, 4,000 words a day. He left the place for the last time in January 1949 but by then it was too late for his health, the die cast. *Nineteen Eighty-Four* was published in June of the year. He was in a sanatorium in Gloucestershire. There were rave reviews.

In his June 1949 review of *Nineteen Eighty-Four*, in the *New Yorker*, Lionel Trilling wrote: 'the final oligarchical revolution of the future, which, once established, could never be escaped or countered, will be made . . . by men of will and intellect.' He concluded: 'Orwell marks a turn in thought; he asks us to consider whether the triumph of certain forces of the mind, in their naked pride and excess, may not produce a state of things far worse than any we have ever known.'

In October Orwell was married to Sonia Brownell, an assistant to his old Etonian friend Cyril Connolly, editor of the literary magazine *Horizon* and possibly the inspiration for Julia, his novel's heroine. His health was ever on the wane and on 21 January 1950 he died of a ruptured blood vessel in one of his lungs.

Perhaps, in his wearied heart, he did not truly leave Jura. He had travelled into the west and never quite left. I like to think that for the duration of his stay on the deer island he was gifted the second sight, with an awful glimpse of what was to come in the twenty-first century. Like all the finest seers, he was dead and buried long before its most poisoned flower could bloom.

Back on Skye, before, after and through the months and years of his writing *Nineteen Eighty-Four* on Jura, the wee black car was coming and going, fast as a shuttle back and forth across a weaver's loom, pulling a thread to make a picture warning of another horror in another place.

When I think of my dad, I try never to see him dead, as I last saw him. Almost always, when he is in my thoughts, he is as he was on that trip to Skye and onwards. Younger than I am now, he is always tall and straight and handsome. I know where to look for that version of him: on Skye. I can find him there.

23

GLENCOE, AND WRONGS UNFORGOTTEN, UNFORGIVEN

'The evil that men do lives after them: the good is oft interred
with their bones.'

WILLIAM SHAKESPEARE, *JULIUS CAESAR*

'May we yield to the [Christ] child so that He may deliver us
from every tribulation. Although He is true to us, there is no
joy without Clan Donald.'

GIOLLA COLUIM MAC AN OLLAIMH,
'THERE IS NO JOY WITHOUT CLAN DONALD'

THERE ARE THOSE WHO WILL TELL YOU GLENCOE MEANS 'THE GLEN OF WEEPING'.
Since the second syllable of the name of that valley in the
northern quarter of Argyll, in Scotland, has no agreed translation, in truth it is likely its roots are deeper even than the Gaelic
language, and so lost. It might mean 'the narrow valley', but there
is no consensus. 'Glen of weeping' is a fantasy indulged by those

of a romantic bent, believing the valley was somehow named in advance of grief to come. By any name it is a dour place, overwhelmingly so, the traveller threatened on all sides by great clenched fists and bare knuckles of rock raised ready for combat. One word that comes inevitably to mind is 'looming'. It is lovely too, a great and terrible beauty.

Most visitors today arrive in Glencoe via the A82 that runs through it, and the impact of the confines of the valley is intensified by what came before, in the south, being the big-sky flatness and lochan-dotted wilderness of Rannoch Moor. The outrider for the grim grandeur of Glencoe, the sentinel guarding the approach, is the frost-shattered pyramid of Buachaille Etive Mor, the Big Shepherd seeming to bear all the weight of the world.

The rock of Glencoe was cooked into existence 400 million years ago when that which would be the Highlands of Scotland was part of an ancient febrile continent south of the equator. Aeons of languid drifting – on rafts of rock floating on Hadean deeps of fire – remained to be completed before it would reach its present temporary, temperate anchorage in the north. Then millions of years of Ice Ages, culminating perhaps ten thousand years ago, saw to the sculpting of Glencoe. A great slab of ice, half a mile thick, bulldozed its way south and scraped the valley into being. Rock that had been scoured clean of life, as though by the hand of a wrathful God, was made a blank slate for the writing of human stories.

Green life came first, though. Trees: birch and Scots pine, alder, elm, hazel, oak. Where there were trees there were soon animals seeking dappled shade. Where deer came, men followed, with spear and arrow. Compared to the age and tortured travails of the rock beneath the young soil, and rearing all around, life was a thin skin lately grown and vulnerable. After hunters, farmers, and pastoralists – herders of cattle. Glencoe became home, in time, to people claiming descent from a long-ago Viking. Somerled – 'led by the summer' – made a home in the west, in the islands.

Centuries passed and some of his descendants, called Macdonalds by then, kept cattle in Glencoe – their own and those they stole from their neighbours. Cattle thievery was hardly the preserve of the Macdonalds alone – a communal Highland pastime rather – but it is fair to say that by the seventeenth century they were among the most notorious.

Notable among the mountains of Glencoe are Beinn Fhada, Gearr Aonach and Aonach Dubh – the Three Sisters – a triptych

Beauty great and terrible . . . the Three Sisters of Glencoe

of summits rising up to the north of the Bidean nam Bian massif. Between Beinn Fhada ('the long mountain') and Gearr Aonach ('the short ridge') lies the hidden valley of Coire Gabhail, which is 'the glen of the capture'. Coire Gabhail – pronounced 'Corrie Gale' – is hidden, or made lost, by the products of an ancient landslide that concealed its entrance and caused a loch to form, like a reservoir behind a natural dam. Beyond the loch a flat pasture is ideal for a clan of cattle thieves in need of somewhere to conceal their stolen beasts.

In Glencoe, as in other places in Scotland's north-west, the world shows her age. Once upon a time the mountains there were mightier by far than the Himalaya. Time has got the better of them, though – the shattered shards that remain are as the ground-down stumps of teeth in an elder's jaw. Bar the weather, and the clouds it conjures, there is little in the way of softening the harsh truth. Where skin is stretched thin over bone, humankind always struggles for life. As often as not, the cruellest threat to that life is not the place, but other people.

Since the death of Good Queen Bess, Stuart kings of Scots had been kings of England too. After the hiatus of Cromwell's Commonwealth, Charles II was restored to the throne of his murdered father, Charles I. And after the death of the second Charles his brother James – II of England and Ireland, VII of Scotland – succeeded him, in 1685. He was Catholic, but even in Protestant England and Scotland his succession was popular enough – until the birth of his son James Francis Edward in 1688. After Cromwell there had been disinclination to take down any more rightful kings by bloody means.

Fears of the establishment of a Catholic dynasty reared, however, like an unquiet ghost. Until little James, the king's heir had been his Protestant daughter, Mary, wife of the Protestant William of Orange. Now Catholic James and his new Catholic heir were altogether less appealing. For William there was Catholic Louis XIV of France to consider too. He had been in the ascendant since 1678 and his triumph in the Franco-Dutch War. What followed was the Nine Years War, in which men wrestled again, like rats in a sack.

Promised a warm welcome by English Protestants, and keen to secure and galvanize support for his struggle with Louis, William brought an army fourteen thousand strong to Brixham, Devon. The royalist army withdrew in the face of him and them, without drawing blood, and men deserting all the while. By Christmas James was in exile in France, and in February, the English Parliament accepted William and Mary as joint monarchs.

In March James was in Ireland, on the comeback trail, securing support from a majority Catholic population by promising restoration of lands and rights. While he laboured on the Emerald Isle, his supporters in Scotland set about stirring an uprising of their own, led by John Graham, Viscount Dundee, *Bonnie Dundee*. There was a pyrrhic victory at Killiecrankie, on 27 July, where Dundee was killed among a heap of government slain, after which support for the rebel cause leaked away.

By early the following year, that first of the Jacobite rebellions was over and the Scottish government was determined to pacify the Highlands. A new garrison fortress was raised at Fort William and more troops deployed at other strategic points. Those men were urgently required for war on the continent, though, and with a view to securing a lasting peace, the Highland chiefs were invited to a parlay at Achallader Castle, at Bridge of Orchy, in Perthshire, in June 1691. The man tasked by the government with overseeing the negotiations was John Campbell of Glenorchy, 1st Earl of Breadalbane, and he was empowered to offer an amnesty to all clan chiefs who had lately sided with James. He had money to offer – bribes totalling twelve thousand pounds that would enable them to secure title to their clan lands. All they had to do was swear allegiance to the new king by 1 January 1692.

The power behind Breadalbane was William's secretary of state in Scotland, and lord advocate, Sir John Dalrymple, the master of Stair. Dalrymple was Williamite all the way down, cold and calculating, and believed that peace in the Highlands would not be secured by mere money. Rather, he was looking to make an example of some luckless clan – and since he was convinced one or other would baulk at the thought of betraying James, he sat back to see who would step into his murderous sights. As it happened, almost all the clans accepted William's deal. Those who refused were led by powerful men in their own right – like Macdonald of Glengarry, secure in a fortified base at Invergarry in the hard-to-get-at Great Glen. Only one chief was weak enough to be made an example of and he – Alasdair MacIain Macdonald, 12th Chief of Glencoe – was no rebel, only a victim of bad luck.

MacIain was bound by an ancient code. He was also a big man, and imposing, well over six feet tall and white-bearded. His word was his bond and, having sworn loyalty to James, he needed the written permission of that king over the water to do otherwise. James understood the predicament of the chiefs, and the necessary letter, letting them off the hook, arrived in the Highlands in late December. That left MacIain but little time to travel from his home in Glencoe to the office of the sheriff at Inverary, the man empowered to take his oath. He made instead for the garrison at Fort William, which was closer, in hope of signing a paper there. It was forlorn. On arrival he was informed Inverary was his only option. Into the teeth of a snowstorm, he set out, detained along the way by government soldiers suspicious of his credentials. After a journey of sixty-odd miles in the worst of conditions he arrived in Inverary on 3 January – two days past the deadline. To make matters as bad as could be, the sheriff, Sir Colin Campbell of Ardkinglass, was elsewhere so it was not until the sixth that he was able to put his name on paper. By the time the document arrived in Edinburgh, the Privy Council refused to add MacIain's name to the list of the chiefs who had sworn allegiance to William. *Alea iacta est* – the die is cast . . .

Dalrymple was cock-a-hoop. Here was what he had hoped for, an isolated section of a clan, militarily weak, in demonstrable defiance of the king, so he moved against them. The Macdonalds of Glencoe had fallen a long way from the glory days of their clan. They remembered when they had stood beside the Bruce at Bannockburn in 1314, but others found it more convenient to forget. Their progenitor Somerled had become Lord of the Isles and ruled a seaborne kingdom to rival that of the King of Scots, but by the seventeenth century those Macdonalds of Glencoe were little more than tenants of their Campbell neighbours. As they raided Campbell cattle, along with everyone else's, there was no love lost between the clans.

King William put his name to the death sentence and it was delivered to the commander-in-chief in Scotland, Thomas Livingstone. Dalrymple selected Campbells for the wet work, in the form of the Argyll regiment. Two captains were tasked: Thomas Drummond and Robert Campbell of Glenlyon, a relative of Breadalbane and a hard-drinking man with gambling debts. Into Glencoe they marched on the pretext of having orders to collect unpaid taxes in Glengarry. They had warrants to billet their men, seventy or so, in the scattered homes of the six hundred and more Macdonalds of Glencoe. Highland courtesy demanded they be made welcome, treated like friends, if not family. For two weeks the redcoats availed themselves of the hospitality of their hosts, eating their meat, sleeping in their beds, keeping warm by their fires and dancing at their ceilidhs.

The order signed by William had been sinister enough. In part it read: 'If MacIain of Glencoe and that tribe can be well separated from the rest, it will be a proper vindication of the publick justice to extirpate that sept of thieves.'

On 12 February, another letter arrived. As it happened, Captain Campbell was hosted that night by sons of MacIain, Alasdair and John. Alasdair's wife was Campbell's niece.

The fresh orders were clear:

You are hereby ordered to fall upon the Rebells, the MacDonalds of Glencoe, and putt all to the sword under seventy. You are to have a special care that the old fox and his sones doe upon no account escape your hands. You are to secure all the avenues that no man escape. This you are to putt in execution at fyve of the clock precisely.

When the hour came the following morning, the Campbells set about their bloody business. MacIain, the old wolf, was shot down as he tried to rise from his bed. In homes up and down, the glen men, women and children were shot, bayoneted or butchered with swords. For all the horror, it was a botched effort. Heavy snow meant companies of soldiers sent to block escape from the valley failed to arrive on time. Hundreds of Macdonalds, some in their nightclothes, fled into the dark and the swirling snow. Both of MacIain's sons escaped, with Alasdair's baby grandson. The younger Alasdair would in time be 14th chief and would fight more redcoats at Culloden, in 1745.

A final death toll is hard to come by. A figure of thirty-eight appears in many books, but the real total may have been in the hundreds. The clan never did reveal the extent of their loss, including those who died of exposure in the hills while soldiers prowled. By any standards it had been an unspeakable and unforgivable horror. While it had been intended to frighten the whole of the Highlands into submission, the Glencoe Massacre had the opposite effect. A simmering fire, hot with hurt and righteous fury, had been set and kindled, and generations to come would blow upon the coals again and again. Such was the outcry that there were, in the end, two investigations into all that had occurred – a first by the Scottish Parliament, and, when that one took no action against the guilty, a royal commission. The latter report was published in full, describing what had happened as an act of treason and murder.

Dalrymple was unrepentant, only furious that any had survived his cull. In the immediate aftermath he had written letters urging the pursuit of those in hiding. He wanted them slaughtered or transported to the plantations. The story goes that Captain Campbell took to drinking even more heavily than before and that during one session in an Edinburgh pub he mislaid the letter ordering the massacre. It duly came into the hands of journalists and pamphleteers, who used it to stir up public outcry. Eventually Dalrymple was driven from office but later rehabilitated, created Earl of Stair in 1703, four years before his own death. Before breathing his last he was instrumental in securing the union of the parliaments of England and Scotland. No one – no soldier, no captain, no advocate, no politician and no king – paid any price for the murder of so many, slaughtered under the orders of their own government.

The memory of the massacre has lingered. For Highlanders it has been the most inexcusable of crimes – murder 'under trust' by people welcomed into homes and treated as honoured friends and guests. The last time I was at the Clachaig Inn, in Glencoe, there was still a sign on the door reading 'No Hawkers or Campbells'. It reads like a joke for tourists, but the barb on the hook is swallowed deep.

There are tales of ghosts and other apparitions in Glencoe, of course. For one thing, folklorists say that night after night before the killings began, the song of the *caoineag*, the 'weeper', of the Macdonalds was carried on the wind. The spirit of a woman wails her lament, often by a waterfall, as a portent of calamity for her clan. In *Carmina Gadelica*, folklorist Alexander Carmichael recorded what she had been heard to sing in Glencoe during those dire weeks of February:

Little 'caoineachag' of the sorrow
Is pouring the tears of her eyes,
Weeping and wailing the fate of Clan Donald.
Alas my grief! that ye did not heed her cries.

HAUNTINGS

There is gloom and grief in the mount of mist,
There is weeping and calling in the mount of mist,
There is death and danger, there is maul and murder,
There is blood spilling in the mount of mist.

Tales were told of those Macdonalds who heard her warning
and took their leave of the glen in time, of Campbells who broke
their swords rather than commit murder. Too many Macdonalds
did not listen, or else took no heed, so it is their ghosts that remain
to remind the living of what took place. When the low light of
winter has gone out of the sky, a visitor might catch a glimpse of
a Macdonald man, or woman, or child, seeking refuge in a fold in
the landscape, in the shadow of a boulder. Sightings are especially
common on the anniversary of the evil deeds – 13 February – and
from time to time there have been those who claimed to catch
glimpses of the slaughter, the burning out of the Macdonald
homes by Campbells bent on erasing the presence of their foe
completely.

Perhaps what is felt most keenly in Glencoe now, by those sensitive to the emotion, is the universal guilt of our species born of
every inhumanity that haunts us for ever, as it should. John Donne
warned against asking for whom the bell tolled – 'any man's death
diminishes me because I am involved in Mankinde ...' There
behind the terrible beauty of Glencoe is the horror of what happened in 1692. Nothing expunges the wrong and therein lies the
sadness that prevails. A perfumed corpse still has the stench of
death upon it, so Glencoe is a glen of weeping now for those with
ears to hear it.

Some of all that happened trails back to King James, the first of the kings over the water and in whose name so many would die in the decades to come. There is a haunting, oddly apposite strangeness about what happened to his body after he died, of a brain haemorrhage, still in exile, at the château in Saint-Germain-en-Laye, in 1701.

Like an Egyptian pharaoh he was eviscerated by his embalmer, his embittered heart boxed in silver and given to the nuns of Chaillot. His intestines were separated into two silvered urns, one for the parish church in Saint-Germain-en-Laye, and the other for the Jesuits of Saint-Omer. His brain was placed inside a lead casket and sent to the Scots College in Paris, and some skin flayed from his right arm was given to the Augustinian nuns of Paris. His hollowed corpse was tightly bound and placed inside a wooden coffin, inside a lead coffin, inside a wooden box draped in black velvet and the whole lot put on display in the Chapel of Saint Edmund in the Church of the English Benedictines in the rue Saint-Jacques in Paris. Candles were set around it and kept lit day after day and year after year.

It seems the Catholic faithful expected that the day would surely come when, with one of his heirs restored to the thrones of England, Scotland and Ireland, his remains would be repatriated for rightful burial among his fellow monarchs in Westminster Abbey. It was not to be, of course, and after many candles had guttered, been replaced and guttered again, the sans-culottes of the French Revolution ransacked his resting place and stripped the lead from his coffin for musket balls. An account of the grisliness was recorded in 1901 in *The Last Words of Distinguished Men and Women (Real and Traditional)* by Frederic Rowland Marvin. According to one Mr Fitzsimmons, an octogenarian Irishman

interviewed in 1840 about what he had witnessed as a prisoner in
that church during the momentous days of revolution:

> ... the sans-culottes broke open the coffins to get at the lead to
> cast into bullets. The body lay exposed nearly a whole day. It was
> swaddled like a mummy, bound tight with garters. The sans-
> culottes took out the body, which had been embalmed. There
> was a strong smell of vinegar and camphor. The corpse was beau-
> tiful and perfect. The hands and nails were very fine. I moved
> and bent every finger. I never saw so fine a set of teeth in my life.
> A young lady, a fellow prisoner, wished much to have a tooth; I
> tried to get one out for her, but could not, they were so firmly
> fixed. The feet also were very beautiful. The face and cheeks were
> just as if he were alive. I rolled his eyes; the eye-balls were per-
> fectly firm under my finger. The French and English prisoners
> gave money to the sans-culottes for showing the body. The trou-
> serless crowd said he was a good sans-culotte, and they were
> going to put him into a hole in the public churchyard like other
> sans-culottes; and he was carried away, but where the body was
> thrown I never heard. King George IV tried all in his power to
> get tidings of the body, but could not.

It is all too macabre for words, glimpses of a younger world.
Jacobus was made into pieces, scattered to the four winds to
linger where he did not belong. Uneasy rests that king's head if it
rests at all. Maybe his spirit is doomed to roam on account of the
hurt that followed in his train. Worth remembering, too, is that
our species has not changed in all the years of our existence on
Earth, or no change worth a tinker's cuss. There are still those,
many, who would keep a relic made of a king's heart, who would

value as a lucky charm a tooth pulled from a royal jawbone. There are always those who would follow orders and commit murder and the rest.

We stand alone on a heathered hillside in Glencoe and glimpse the spirit of a murdered child, or a mother or father in restless search of those they have lost. The ghosts are made of what we know happened and what we know to be wrong. What was wrong before is wrong for ever; our hearts are broken and remain so. Pain alters our view – what we see and how we see it – and we have learned to call it ghosts.

24

Tigguo Cobauc, and the Fear and Lure of the Dark

'I, the star god,
take bones from the
underworlds of past times
to create mankind'

ANONYMOUS, QUOTED BY HANNAH
GAMBLE IN THE POETRY FOUNDATION

FOR AS LONG AS OUR SPECIES HAS BEEN UPON THE EARTH, WE HAVE BURROWED into the rock of it. Or where water went first, carving tunnels and chambers, we followed in our own time, into the dark. It is an urge primal and irresistible, or else a homecoming. Out of the womb of the mother comes man, and back into the womb of the world he crawls in search of safety out of sight. All around the globe, however deep, cavers and spelunkers have crawled and squirmed. With bottled air to breathe and electric lamps to light the way, they have found the bones of ancient fellow

291

travellers, precious things left behind by them, their art works daubed on walls.

We are also frightened down there in the dark, frightened of the devouring mother, and in those caves, in that darkness, creatures waited for our predecessors: spiders big as rats, scorpions with stings raised. Our fear of things that scuttle, things that bite, was born in dark places. For the longest time any cave our ancestors might access likely had great bears or sabre-toothed cats as sitting tenants. Some of the oldest remains of our distant ancestors have been found in caves – skulls and long bones marked by the teeth of the predators that dragged them there for the eating. Worst of all, even caves devoid of animal life contained the dark – the reservoir of deepest fears. In the dark spaces all the ghosts of the world are patiently waiting.

Beneath the city of Nottingham, in the middle of England, is another hidden conurbation, of caves, home and refuge for locals for the better part of two millennia. They have been on the record since the ninth century – when Asser, Bishop of Sherborne, wrote about finding people living in what he called Tigguo Cobauc, Old Brythonic for 'the place of the caves' – but archaeologists have established that some of the earliest tunnelling into soft sandstone dates from the AD 300s.

By now the total of explored caves numbers more than five hundred, but others may yet remain to be rediscovered. Rather than any work of nature, every space beneath Nottingham has been carved out by people into homes, stores and tanneries, air-raid shelters and cesspits. Always and in every way, they have cradled life out of sight and out of reach of the world above. Scores of Nottingham pubs have cave systems beneath them – used as overspill accommodation for drinkers and as cellars for storing beer and food.

The Bell Inn on Angel Row was built as a refectory for the Carmelite monks resident in the monastery on Beastmarket Hill. Timbers within reveal, via their rings, that some of the structure dates to 1420 or thereabouts. After Henry VIII took the monasteries and their lands for himself, the place was made into a public house.

Angel Row and Bell Inn take their names from the Angelus bell that was once in place nearby and rung to call the faithful to prayer, its tolling reminding all of how Christ was made flesh in the womb of Mary ... *Angelus Domini nuntiavit Maria* – 'The Angel of the Lord declared unto Mary'. Beneath the Bell Inn a cave system spreads under adjacent properties and makes space too for two wells the monks used for brewing their beer. The Bell is hardly the only pub in the city claiming great age: Ye Olde Salutation Inn at Hounds Gate has roots twisting down to 1240, when it was called the Archangel Gabriel Salutes the Virgin Mary – more of the same devotion recalled by the Bell. The caves beneath the Sal, as locals know it, are larger than most: cellars with rock-cut shelves for the safe storage of food, a well cut eighty feet into the rock.

Nottingham Castle perches atop more of the same soft rock. The building there now is seventeenth century but squats on the site of a castle built by Normans in the eleventh century, which had multiple iterations in the centuries between. From within the walls a three-hundred-step passageway winds its way down to another pub vying for the title of oldest in England – Ye Olde Trip to Jerusalem, in Brewery Yard. Its claim for oldest of all cites 1189 as the year of its creation – the same that saw Richard the

Lionheart crowned king and Pope Gregory VIII call for the Third Crusade.

The tunnel is called Mortimer's Hole and named for Sir Roger of that name who, with his lover Queen Isabella, the She-Wolf of France, murdered her husband the king, Edward II, and seized his throne for themselves. Perhaps it ought better to be named Ned's Hole, since it was via the tunnel that young Edward III and his men gained entry to the castle on 19 October 1330 and laid hands on the traitors. Sir Roger took the drop at the Tower of London while his conniving lover was let off with enforced retirement. *Noblesse oblige.* As well as Mortimer's Hole, the Trip is adjacent to yet more caves burrowed out of the castle rock – and, like the Sal, is said to have hosted crusaders preparing for the long, thirsty voyage to the Holy Land.

While Mary King's Close in Edinburgh has little orphan Annie, the Sal has Rosie, another ghost of a girl. Like Annie, Rosie does not want for toys, at least, since regulars and visitors alike are wont to leave dolls and such for her entertainment in the dark when all have gone. Underground has always been the place of the dead. Our lives in the world above might be followed by an eternity below. As poet Roger McGough had it, our mortal remains, our 'left-overs', are boxed up and 'lowered into a space on loan from the clay'. The thought of all those buried dead must preoccupy the living.

Underground, underworld, words laden with mystery and foreboding, abodes of the dead, and the gods that rule them. Places to enter, never to return. Tigguo Cobauc was burrowed out of the earth for practical reasons: by people seeking shelter from the elements above. All around the world the same need prompted more burrowing, more borrowing of space from the

The Troops of Lord Montacute in the Subterranean Passage.

Treason undone . . . access to Nottingham Castle for troops loyal to Edward III, via Mortimer's Hole

bedrock. In Cappadocia, in central Turkey, people of the Phrygian culture began digging down into soft volcanic rock sometime in the Bronze Age.

By Roman times, the tunnels and caves they left had been acquired by Christians in need of safety from persecution, and they expanded the earlier workings into a multi-level city. Known now as Derinkuyu, the vast subterranean complex reaches, at its deepest level, nearly three hundred feet below the surface. Air shafts make possible living spaces, stores, communal chambers, stables for animals and chapels with vaulted ceilings. There is a cruciform church. It is estimated that as many as twenty thousand people may have been accommodated in Derinkuyu for extended periods. Tunnels leading to the surface could be blocked against attackers with huge circular stones. Wells provided water. Abandoned in the 1920s, the underground city was rediscovered in the 1960s and is in use again for all manner of storage.

But always the underground, the underworld, haunts our thinking. In Greek myth there is *katabasis*, a journey undertaken by a hero into the Underworld, into the realm of the dead, in search of something or someone lost and wanted. Odysseus enters Hades and encounters souls there, including that of his mother and also Tiresias, the Theban Seer. Theseus entered the darkness of the labyrinth to slay the Minotaur. The Welsh hero Pwyll, Prince of Dyfed, swapped places with Arawn, ruler of the Otherworld of Annwn, for a year. In the Harrowing of Hell, Jesus descended into the Underworld for the time between his crucifixion and his resurrection: 'For this cause was the gospel preached also to them that are dead ...' And for the Maya of Meso-America there were the Hero Twins – Hun Ajaw and Yax Bahlam – who were summoned to Xibalba, the underground Place of Fright, to play the Ball Game against the gods.

In Norfolk, four and a half thousand years ago, farmers turned miners and opened great shafts into the earth in search of the deeply buried flint from which they made their tools. Once each opening had surrendered all it had to give of the brittle, glassy stone, it was backfilled – but not before offerings were made and left in the dark, thanksgiving to the ever-present spirit of the world. Long after, Anglo-Saxon farmers found the pitted land-scape and called it Grime's Graves – by which they meant the quarries of Grim, who is the hooded man, or Woden.

Some of those mines have been opened again, emptied by archae-ologists. To descend now into the half-light of the chambers below is to enter another borrowed space that feels home still to that world of four thousand years ago. Picks made of red-deer antler stand propped against the walls, where they were set by those long-ago miners at the end of a long-forgotten day, and the cool still air feels woven and stirred by their presence, as though they had climbed out and walked away just moments before.

Hollowed-out spaces – man-made and natural – are imbued and impregnated with essences of the past that linger and coil like undisturbed smoke. In the commune of Montesquieu-Avantes in the central Pyrenees in south-west France, tunnels, caves and chambers were carved over millennia by the River Volp. Three systems, like three wombs, lie side by side. To the east is Enlène Cave, in the middle the Cave of Les Trois Frères – the three brothers – and in the west the Tuc d'Audoubert. Les Trois Frères and Enlène are interconnected, but Tuc d'Audoubert stands alone. All three were used by artists long ago as the canvas for paintings.

Enlène has the fewest pictographs and petroglyphs while the other two are awash with a cavalcade of animals, a veritable torrent of creation flowing over walls and ceilings. The ghosts of the artists who swam, crawled and wriggled their way into the dark, seem all around. Their fingerprints are in the ochre and other pigments with which they conjured their works into being.

And who are we, anyway, to apply a word like 'art' and imagine we understand why those images were made and left behind fifteen thousand years ago? Perhaps those spreads of hands, dots and engravings, the black lioness, birds and cats, horses, reindeer and bison, bears, goats and mammoths, the image of a half-man half-deer called the Sorcerer were prayers to the spirits of all – or something else we lack the wit to divine. All those thousands of years before Plato, were they evocations, in the secretive, protective dark, of the 'forms' of fellow travellers, their true natures?

In the Tuc d'Audoubert there are nearly four hundred separate images. Some were found during the eighteenth and nineteenth centuries, but it was in 1912 that the landowner, Count Bégouën, and his three sons – the eponymous three brothers – explored the systems using a boat they had built for the purpose. By the light of lanterns, they poled their craft further and further into the dark until they reached a broad chamber and saw a narrow opening in one wall. The brothers were able to wriggle through quite easily but then the hole had to be widened, chipped and smashed to let their father follow. What they found, in the deepest reaches of the Tuc d'Audoubert, has mesmerized the art world ever since: two bison, a male and a female, about to mate, carved in relief in clay scraped from the walls of the cave. The figures are around eighteen inches high by perhaps two feet long and standing proud of the rock upon which they were made by perhaps four inches. The surface of each is smooth, as though still wet from the artist's finishing touches. Some sort of stencil was used for the making of lifelike manes and beards on each, but the artist used his or her fingernail to realize the shapes of their snouts and jaws. The footprints of the artist and those present while the work was done are still there on the cave's clay floor, surely proof of ghosts if ever there was.

Writer and subterranean explorer Will Hunter, author of *Underground: A Human History of the Worlds Beneath Our Feet*, has described the mix of terror and elation to be found in the realm down under, and suggests it has ancient roots:

> As we evolved for life on the African savannah, where we hunted and foraged in daylight and where nocturnal predators stalked us in the night, darkness unnerved us. But the subterranean dark – 'the sightless world' as Dante called it – is enough to cause our entire nervous system to splinter.

He wrote of how 'prolonged immersion in absolute darkness can trigger psychological aberrations, a phenomenon that cavers have labelled 'the rapture'.

> Down in any underground hollow we feel, if not the full tempest of panic, a reflexive tingle of not-quite-rightness, as we imagine ceilings and walls closing in on us. But ultimately, of course, it's death we fear most. All of our aversions to the dark zone come together in the dread of our own mortality.

Back in Nottingham the caves lie quiet and forgotten for the most part. Like the rest of the city's history, they are out of sight of most. In

medieval days a bustling area by the River Trent came by the name
of Broad Marsh, and for a time housed a Franciscan priory called
Greyfriars, and its adjoining graveyard. The churchmen lost their
hold on the place in 1539, during King Henry's dissolution of such,
and all too soon it was reduced to a sprawling slum. It was not until
the 1970s that the area was cleared for the building of Broadmarsh
Shopping Centre – and it was while the foundations were being dug
that the caves lying beneath all were revealed once more. Now visitors in search of the underground experience may access some of
what remains via the modern complex. Even so, the mass of the
city's population is largely unaware of what lies beneath their feet,
ghosts and all. Most have never gone below, and in the hundreds of
empty spaces, only silent darkness endures.

In the West we are becoming cave dwellers once more, one way
and another. Technology makes it easy to stay at home behind
closed doors and drawn curtains. We call it working from home,
but it's also about hiding from the outside world. We no longer
have to speak to others, communicating instead via apps, texts
and emails, holding them at arm's length. The future looks lonely,
and also quiet. Instead of the warmth of the sun our faces bathe
in artificial light, in the glow of smart phones and computer
screens. The Fourth Industrial Revolution is under way and
making us, even more than before, servants of machines.
Uncounted millions are demonstrably anxious and unhappy;
young and old are rattling with sedatives and anti-depressants to
help them cope with lives made insular and isolated. We are with-
drawing, retreating into caves of our own creation. Nobody and
nothing waits for us there, except ourselves.

25

NUMBER 50 BERKELEY SQUARE

'If there be physiognomy in bricks and mortar, one would say that house has seen murder done ... a valuable house left seemingly to decay, with windows caked and blackened by dust, and yet with no notice about it anywhere that it may be had for the renting. This is known as "the haunted house in Berkeley Square".'

MAYFAIR MAGAZINE, 1879

'The stories about the ghost are complete horlicks ... Nothing to it at all.'

MAGGS BROTHERS LTD

EVERY HOUSE IS HAUNTED IF YOU LET IT BE ... WANT IT TO BE. IN 1980 MY FAMILY moved into a house on Victoria Terrace, in Dumfries, a market town in south-west Scotland. A plaque on the red sandstone wall declares the author and playwright J. M. Barrie lived there between 1873 and 1878, 'while a scholar at Dumfries Academy'. It also describes Dumfries as 'the town where Peter Pan was born'. I was thirteen on arrival at that address and remember being faintly excited by the idea of rooms once occupied by a named person from the past, a writer whose most famous book I already had in my possession. What had the house been like when he walked its floors, what pictures on the walls? What bedroom had been his? I attended Dumfries Academy as well and wondered if he had taken the same route to school as I did, past the same railway station, the same church, the same Victorian and Edwardian houses.

Barrie was born in Kirriemuir, in Angus, in 1860, the ninth of ten children. Two of his siblings died before he was born. One of his elder brothers was David, handsome and clever, their mother's favourite. In the winter of 1867, when the lad was thirteen, he went ice skating. There was an accidental clash of heads and David, one day short of his fourteenth birthday, suffered a fractured skull and died. The grief of their mother, Margaret Ogilvy, sent her to her bed for weeks. The author wrote later about going in to see her:

The room was dark, and when I heard the door shut and no sound come from the bed I was afraid, and I stood still. I suppose I was breathing hard, or perhaps I was crying, for after a time I heard a listless voice that had never been listless before say, 'Is that you?' I think the tone hurt me, for I made no answer, and then the voice said more anxiously, 'Is that you?' again. I thought it was the dead boy she was speaking to, and I said in a little lonely voice, 'No, it's no' him, it's just me.' Then I heard a cry, and my mother turned in bed, and though it was dark I knew that she was holding out her arms.

Loss and grief entered Barrie's imagination then, twining and tying in a hard and confusing knot. It seemed to him that the lost boy was made perfect in his grieving mother's mind, left untouched by the passing of time, set beyond the reach of messy adulthood and all its failures and disappointments. In the years following David's death, Barrie endeavoured, for his mother's sake, to replace him – mimicking his whistle and his ways, on one occasion dressing in his dead brother's clothes. Perhaps in the end it seemed to him better to avoid growing up altogether. Barrie was always small, only a little over five feet and three inches tall in adulthood. He favoured overcoats that drowned him, as though he were a child wearing his father's clothes.

Margaret Ogilvy had been eight years old when her own mother died and she was cast, too early, into the role of mothering her younger brother. Like Peter Pan's Wendy, she was a child tasked with motherhood before her time. Her grief at the loss of David stayed with her, with the family, fixed in childhood, as a leaf in amber, his absence ever present. She remained somehow absent too, and Barrie's elder sister, Jane Ann, had to step in as a surrogate mother. (One of my elder sisters is called Jane-Ann and lives in the house still.)

Once Barrie started writing, his preoccupation with lost boys and never growing up was a recurring theme. While living in Kensington, in London, he befriended the Llewelyn Davies family, Sylvia and Arthur and their two boys, David and Jack, aged five and four. Eventually there would be three more sons – Peter, Michael and Nicholas – and Barrie became Uncle Jim to all five. Cynics have sought since to find something sinister in the affection he showed the boys (his own childless marriage failed while the relationship with the Llewelyn Davies family lasted until his death in 1937) but it seems more likely he wanted only to eavesdrop on happier childhoods than his own. He may have been ghost-hunting too, consciously or unconsciously on the lookout, in the lives of others, for evidence of the little boy his mother lost.

All of his borrowings, vicarious living and reliving through other children's lives had its brightest flowering in *Peter Pan*, of course – a story he wanted to call 'The Boy Who Hated Mothers' until his publisher talked him out of it. In fact, the novel is called *Peter and Wendy* and was first published in 1911. Before that, the character of Peter Pan had featured in other of Barrie's stories and in the stage play. Peter Pan had run away from his own mother when he sensed her wanting him to begin the tiresome business of growing up and becoming a man. He preferred Neverland and the company of other perpetual little boys. When at last he went in search of a little girl to mother him and them, he was ultimately betrayed when Wendy grew up and became a mother in her own right. Towards the end of the story, he finds her again and neither he nor she can cope with her womanhood:

> He was a little boy, and she was grown-up. She huddled by the fire not daring to move, helpless and guilty, a big woman.
>
> 'Hullo, Wendy,' he said, not noticing any difference, for he was thinking chiefly of himself; and in the dim light her white dress might have been the nightgown in which he had seen her first.
>
> 'Hullo, Peter,' she replied faintly, squeezing herself as small as possible. Something inside of her was crying, 'Woman, woman, let go of me.'

Wherever Barrie lived (in that house in Dumfries, as well), did he bring with him his brother's ghost? Did he hope, like Peter Pan with his shadow, to find and stitch back on to him that which was otherwise missing? Wherever there are houses with long lives and many stories, there are ghosts, one way or another. The house in Stirling where my family lives now was home early on to the Reverend Ebenezer MacLean and his wife, Elizabeth. Elizabeth was a suffragist – demanding votes for women but eschewing the physical violence applied by suffragettes – and a campaigner on behalf of the British Women's Temperance Society. She knew the Liberal Prime Minister Henry Campbell-Bannerman and the suffragette leader, Emmeline Pankhurst. We bought the pile from a retired veterinarian who lived there with his family, the Lovetts, for fifty years. They had bought the house the same year I was born. Part of the building housed his surgery and waiting room and we have yet to get round to using them as more than storage and workspaces. The waiting-room sign is still on the door, labels for drugs and other chemicals – sulphonamide, nembutal – still stuck on the shelves. Fingerprints of ghosts of other lives.

Last year Mr and Mrs Lovett died in grand old age. Mr Lovett's funeral service was held in the church just a hundred yards from his old front door. Not long after we moved in, one of our friends paid a visit. She says she is a medium, and who are we to doubt her? After my wife showed her round the old surgery, she said she had seen an old man seated in the waiting room, smoking a pipe. My daughter's bedroom is on the top floor where, in bygone days, the children's nursery was located. In the surgery there's a box on the wall that was once the hub of the system of bells that summoned servants long ago: Hall, Dining Room, Waiting Room,

Drawing Room, Bathroom, Bedrooms 1, 2, 3 and 4, Side Door, Day Nursery ... Evie says she has, more than once, felt hands placed gently on her shoulders while she sits on her couch, a steady pressure. She senses no malice whatever, quite the contrary, in fact. I don't doubt she's telling the truth. I am writing a book about ghosts in a house that seems full of them.

In the house where I lived with my parents and my sister (that second Jane-Ann to inhabit the place) my dad used to tell us he glimpsed, from time to time, the movement of 'little people', like Borrowers or fairies. It always happened while we were sitting in the living room, watching television or reading. Out of the blue Dad would suddenly point towards the skirting board and say, 'There! Did you see them, that time?' No one ever did, but he was ready to swear on a stack of Bibles that he wasn't making it up. I used to wonder if Tinkerbell and her fairy ilk were not occasional visitors to the house where Peter Pan was born, summoned as though by bells.

During part of the nineteenth, all of the twentieth and still now in the twenty-first century, the house at 50 Berkeley Square, in London, had a reputation as 'most haunted'. Until 2015 it was occupied by antiquarian booksellers Maggs Bros Ltd, founded in 1853 by the Dickensian-sounding Uriah Maggs. Changing hands within, over the years, were the fifth-century Codex Sinaiticus (being the Sinai Bible, containing all of the New Testament and some of the Old, handwritten in Greek); not one, but two Gutenberg Bibles; Caxton's first book printed in England; and the penis of Napoleon Bonaparte. In such ways, by such unlikely events, are places imbued with that sense of something strange. Just as the presence of King James's intestines or flayed skin make odd and inexplicable the atmosphere inside the French church

London's most haunted . . . 50 Berkeley Square

buildings in which they were secreted, so the stay of Bonaparte's member, however brief, within the walls of 50 Berkeley Square is enough to set the place apart all on its own. Author John Vernon followed the unlikely trail of the tale of the emperor's tail while researching a novel about it. He wrote:

> In a long footnote in a book on Napoleon's death by the Scottish physician Frank Richardson, I found at last what I was looking for, a discreet and skeptical reference to the story of this Promethean theft. The item was mounted in a velvet Cartier box and sold by the descendants of Abbé Ange Paul Vignali (Napoleon's chaplain, who was present at his autopsy) to the Maggs Brothers of Berkeley Square in London. (*New York Times*, 12 July 1992)

Of all the gin-joints in all the world, it was that particular address in England's capital that was made stranger still by such an association. Vernon wrote:

> Those with a passionate interest in the past always feel sooner or later that history's ultimate unknowability mocks them. This could be just paranoia, but it is more: it is the powerful feeling that history is simultaneously there and not there, real and illusory – a ghost forever trailing behind, which vanishes when we turn around.

What is it about some places and not others, about one location among the thousands, the tens of thousands, in a city that earns a reputation for being haunted and haunting? How and why does the strangeness start? The house in Berkeley Square, designed by architect William Kent, was built in 1750. For many years Tory statesman George Canning (until Liz Truss, the briefest-serving British prime minister, with a tally of 119 days, cut short by his death in 1827, compared to her 49) had it as a home. The Honourable Elizabeth Curzon was next, lasting in place until she died in 1859, aged ninety-one. After Elizabeth the house was acquired by a young man in possession of good prospects and in want (and great expectation) of a wife. His name was Thomas Myers, the son of another Member of Parliament, and he took on the address after becoming engaged. According to a memoir written by his relative Lady Dorothy Nevill, young Myers 'made every preparation to receive his bride in it – ordered carpets, pictures, china, everything'. A few days before the nuptials, however, his intended informed him she had decided to marry another. And Myers, wrote Nevill, 'remained there, leaving everything in exactly the same state as when he heard the news which had ruined his life'.

By Nevill's estimate, Myers, an eccentric to begin with, suffered a full-blown breakdown in the aftermath of his rejection. He neglected the house in every way, leaving its exterior to become steadily begrimed, wreathed in the city's pollution as well as with his misery. He retreated to a small room at the top of the place and became nocturnal in his habits, roaming the darkened hallways and lighting his way with a single candle. Neighbours and passers-by would see the trailing, flickering light, will-o'-the-wisp, behind windows caked with dust and dirt, and wondered who or what lurked within. He was given to mutterings and moaning. He never left the house, or not if he could avoid it, so that in time it was forgotten that anyone lived there at all – it must be an empty house, inhabited only by memories and the invisible, as Walter de la Mare notes in 'The Listeners':

But only a host of phantom listeners . . .
Stood listening in the quiet of the moonlight . . .
Stood thronging the faint moonbeams on the dark stair,
That goes down to the empty hall . . .

Dickensian Uriah Maggs and his books would come later, but it is surely fair to suggest poor Mr Myers might have inspired the characteristics of the novelist's jilted bride, Miss Havisham, in *Great Expectations*, living surrounded by an eternity made of the moment in which her heart was broken: 'I saw that the bride within the bridal dress had withered like the dress, and like the flowers,' notes Pip, 'and had no brightness left but the brightness of her sunken eyes.'

Like a ship lost, Miss Havisham, let down on her wedding day, has slipped any mooring in the land of the sane. But she has endeavoured most firmly to anchor herself in time, stopping all timepieces at the very moment when she heard the dread news.

It was when I stood before her, avoiding her eyes, that I took note of the surrounding objects in detail, and saw that her watch had stopped at twenty minutes to nine, and that a clock in the room had stopped at twenty minutes to nine.

'Look at me,' said Miss Havisham. 'You are not afraid of a woman who has never seen the sun since you were born?' . . .

It was then I began to understand that everything in the room had stopped, like the watch and the clock, a long time ago . . . Without this arrest of everything, this standing still of all the pale decayed objects, not even the withered bridal dress on the collapsed form could have looked so like grave-clothes, or the long veil so like a shroud.

Myers lived at number 50 from 1859. His hermit-like existence was not disturbed, it would appear, until 1873 when he was briefly the subject of an investigation about unpaid taxes. According to lore, the magistrate took pity on the man who lived 'in the haunted house' and closed the case without demanding so much as a penny. Myers died soon after, and while his sister inherited the place, it seems she left it empty. Now quiet altogether, the house was made home instead to the imaginings of those who only looked at it or walked past and noted its decrepitude, so out of keeping in such a wealthy locale.

In his 1913 offering, *Haunted Houses*, Charles Harper struck the recurrent note of the publications of the time:

The haunted house in Berkeley Square was long one of those things that no country cousin come up from the provinces to London on sight-seeing bent, ever willingly missed . . . But truth to tell, its exterior is now a trifle disappointing to the casual seeker after horrors. Viewed in the afternoon sunshine with a milkman delivering the usual half-pint, or quart, as the case may be, it is just as respectably commonplace as any other house of similar late Georgian period, and even at the weird stroke of 12, when the midnight policeman comes and thrusts a burly shoulder against the front door, and tries the area-gate or flashes a gleam over the kitchen windows from his bull's-eye, there is nothing at all hair-raising about it . . .

But he also carefully recalled the long years

when No. 50 . . . wore an exceedingly uncared-for appearance. Soap, paint and whitewash were unused for years, and grime clung to brickwork and windows alike. The area was choked with wasted handbills, wisps of straw, and all the accumulations that speedily made a derelict London house. The very picture of misery; and every passing stranger stopped the first errand-boy, and asked various questions, to which the answer was, generally, "Aunted 'ouse'; or, if the question happened to be 'Who lives there?' the obvious answer was 'Ghostesses.'

With Myers forgotten, after the long decades of his withdrawal from society, darkly excited talk had filled the spaces left behind. Accreting around the address, like limpets upon a shipwrecked hull, there were plenty of tales of 'ghostesses'. Writers latched on to the spoken word, stoking and sharing all manner of tales. Much of the unease was reported to centre on an attic room (Myers's bed-room?) where, it was said, a young woman had killed herself by leaping from the window. Depending on which accounts a person reads, her essence is either a figure glowing white, or a drifting brown mist. There is talk too of the ghost of a murdered child, victim of a servant, and of a young man held there as a prisoner of some sort and fed through a hatch in the door until he succumbed to madness and death. From time to time a pair of legs has been seen emerging from one of the chimneys; there has been talk of a 'feathered thing' and a 'nameless horror'. Harper perhaps caught the best of the sense of it with his report of an 'unnamed Raw Head and Bloody Bones'. Over the years – at least according to the reporters of the penny-dreadfuls that fuelled the myth – there were those who sought, for their own reasons, to stay in the house, in the attic bedroom, to confront the phantoms.

An edition of *Mayfair Magazine* in 1879 gave a typical summary of events: the man who rented the place for himself and his daughters, one of whom complained of a heavy scent in the house she likened to that of animals in a zoo. The fiancé of one of the girls, a Captain Kentfield, made arrangements to stay so a maid was despatched to ready his room. All too soon her screams were heard and, when others ran to her aid, they found her curled in a ball on the floor, muttering, 'Don't let it touch me.' In suitably melodramatic style the poor soul was dead the next day. The captain was as good as his word and embarked upon his vigil. Climbing to the room by candlelight he was inside for just half an hour before blood-curdling cries were heard. Once more a party raced to offer help and found the poor man dead on the floor, his face a mask of horror . . .

Other accounts tell of a party of sailors who broke into the place and died soon after; of an unnamed man who took a bell with him into the room, having left instructions that if he were in need of help, he would ring twice. In the small hours, the bell was heard ringing wildly and his would-be rescuers found him having a seizure. Like Captain Kentfield and the rest, he succumbed to his terrors and died without ever managing to recount a word of what he had seen, or what had seen him.

Although *Great Expectations* was written elsewhere, Charles Dickens wrote *The Pickwick Papers* while seated at one of the tables in Maggs Bros. The company moved out of Berkeley Square in 2015 and renovators moved in to turn the place back into a private house. All talk of ghosts there has fallen away.

The floors of my childhood home in Dumfries were walked by J. M. Barrie in the years before his own stories came to him. He was haunted in his own way and all his life by the memory of his perfect brother, David, the little boy who never grew up. My dad said he saw the little people there, the fairy folk. Our daughter says the top floor of our house in Stirling is home still to the presence of a nanny who cared for children there. In our mothballed waiting room, our friend smelt pipe smoke exhaled by an elderly gent still waiting. I am at home to all of it. Why wouldn't I be?

One way or another, 50 Berkeley Square was long a place of stories, some whispered in the drawing rooms of neighbouring houses, some detailed in smudged print in gossip sheets and rags. For seventy-eight years it was a refuge for the stories held inside second-hand books neatly stacked. Perhaps the building, all five storeys of it, was always and only about stories.

In his novel *Little, Big,* John Crowley wrote about a house called Edgewood that was a doorway to other worlds, including the one that is home to fairies, that unseelie court. By the end, its inhabitants have all moved on, leaving the empty house to slip into elegant decline:

> One by one the bulbs burned out, like long lives come to their expected ends. Then there was a dark house made once of time, made now of weather, and harder to find; impossible to find and not even as easy to dream of as when it was alight. Stories last longer, but only by becoming only stories.

Where truth and reality are absent, make-believe might fill the space. Whether anyone will ever again hear the mutterings and moans of Mr Myers, or glimpse the shining presence of the murdered girl, or the nameless shadow, or the feathered thing . . . only time will tell.

26

RAVENSER ODD AND FOR WHOM THE BELL TOLLS

'We must build dykes of courage to hold back the flood of fear.'

MARTIN LUTHER KING JR

'I call it relief, though it was only the relief that a snap brings to a strain or the burst of a thunderstorm to a day of suffocation. It was at least change, and it came with a rush.'

HENRY JAMES, *THE TURN OF THE SCREW*

EVEN ON THE BRIGHTEST DAY, WHEN ALL SEEMS RIGHT WITH THE WORLD, THE unthinkable is there at the back of our minds, the monster ghosting in our peripheral vision. It is the thought – the thought we keep at bay as best we can – that at any moment the worst might happen, the end of our world might appear out of a clear blue sky. In Hinduism that worst is manifest in Kali who, as well as being the Mother Goddess, is also the Devouring Mother, the

Goddess of Destruction, who wipes away the whole world at the end of time. No matter what we have, no matter how dearly loved and tightly clasped, we perceive that we might lose it. Life is short, and also fragile, and we know it. This is not idle fretting: all around us in the landscape is proof that nothing lasts for ever. There are ghosts everywhere – ghosts of ways of life, ghost villages lying in shallow graves, bones protruding. Sometimes we drown our own.

In 2018 I wrote *The Story of the British Isles in 100 Places*. I wrote it out of love for the place I grew up, and also as eulogy for something I felt was dead already. The British Isles – or however they might be named – are still here. The dry land is still mostly proud of the sea. But one idea of the place has been replaced by another. And so it goes, as Kurt Vonnegut said.

The port of Ravenser Odd was founded in 1235 at the tip of a sandbank at the mouth of the Humber estuary. That most marvellous name has its roots in Old Norse, and Hrafn's Eyr, which means 'the tongue of the raven' and so described the elongated spur on which it was presumptuously placed, in defiance of the sea. By the last year of the thirteenth century, when it was made a borough, there were a hundred houses. There was a seawall and a harbour, wharves and warehouses, the needful things of a port. There was a courthouse and a prison and, all in all, Hrafn's Eyr or Ravenser Odd, so ideally placed for trade with the international trading bloc of the Hanseatic League, was counted bigger and more important than nearby Kingston-upon-Hull.

Pride comes before a fall, and for all that life might have seemed good, for some few generations of merchants and their families, the worst was waiting, as it does. Then as now, the soft sand and sediment of England's east coast was easy prey for the sea. By the middle years of the fourteenth century, erosion had taken its toll. The port was flooded again and again. The brief glory days had passed and the enervating toll of abandonment was under way.

The end came in the middle of January 1362. Ireland felt the storm first. Dublin was devastated. Southern England was next, thousands of trees flattened as though stamped on by a wrathful goddess. Impertinent spires on churches and cathedrals were toppled. When the low pressure arrived over the North Sea it was in time to join forces with an especially high tide born of a full moon. The resultant storm surge wreaked havoc not just on England's east coast but all around the North Sea – in Denmark, Germany and the Low Countries. Accurate numbers are hard to come by, but around twenty-five thousand people are thought to have lost their lives. The event is remembered either as the Grote Mandrenke – the Great Drowning of Men – or St Marcellus's Flood, since the worst of it took place on his feast day of 16 January. (In fact, it is more precisely labelled the *Second* St Marcellus's Flood since the same coincidence of events on the same day in 1219 left perhaps thirty-six thousand dead in the Low Countries.) By the time the great drowning was over, that which had been Ravenser Odd, however diminished it might have been by years of encroachment, was utterly gone, submerged beneath the sea.

All around the coastline of Britain there are similar stories of loss, too many to count, that which was briefly ours taken back by the world when it suited. The most famous of Stone Age villages, Skara Brae, on Orkney, was swallowed by other storms in the years before the Pyramids were built in Egypt. First established around 3200 BC, Skara Brae was home to scores of people for perhaps seven hundred years. It was a world of odours – a smelly Hobbiton of semi-subterranean homes packed all about with a midden, the detritus of generations, so that in the end they lived insulated within their own rubbish heap, ripe and warm.

When the living began there, the farmers would have squinted at the sea a mile distant. By now Skara Brae is lapped by waves, testament to the insinuation of the sea during those centuries and in the millennia since. The loss of the village may have been attritional, winter by winter the storms edging closer. Or perhaps a single catastrophic event forced families to flee, surrendering their homes to forces too great to be withstood any longer.

At the other end of the long island of Britain is Dunwich, in Suffolk, another ghost town, the poorest shade of its former self left haunting the tattered edge of its old demesne. Novelist Henry James visited what remained in 1897 – two years after the Archbishop of Canterbury told him the story he was shortly to start writing as *The Turn of the Screw*, a story that existed then only as notes: 'the story of the young children . . . left to the care of servants in an old country house through the death . . . of parents . . . The servants die and their apparitions, figures, return to haunt the house and children, to whom they seem to beckon . . .'

So he had ghosts on his mind when he roamed what remained of a town that had once rivalled London in size and significance. Perhaps founded as early as the seventh century, as the capital of the kingdom of the East Angles, it grew and prospered until the thirteenth and the viciousness of the same decades of depredation that did for Ravenser Odd. Taken, too, by the Grote Mandrenke, it was left a rump. All eight of its churches went. When James visited, only the wreck of the thirteenth-century friary of All Saints was still standing. 'Dunwich is not even the ghost of its dead self . . .' he wrote. 'All the grossness of its positive life is now at the bottom of the German Ocean, which moves for ever, like a ruminating beast, an insatiable, indefatigable lip.'

At All Saints he saw the erosion of its graveyard: 'The bones . . . protruded gruesomely from the cliff, and a single gravestone, to John Brinkley Easey, stood in an inconceivably bleak loneliness at the cliff top.'

We might trick ourselves into thinking that familiar places in the landscape, made of centuries and longer, will always be there. The experience of our species, however, is that destruction waits patiently. Kali is She Who Is Black or She Who Is Death. She is depicted naked, black or dark blue. Her tongue lolls long, a

A landscape of loss and lost . . . Dunwich All Saints

raven's tongue, you might say. She has four arms, or eight, or more. She wears a necklace of severed heads and holds another, by the hair, in one of her hands. Often, she is shown dancing on the prostrate body of her husband, Shiva.

Kali belongs in the East but she embodies the same end of the world we all dread. Our species has feared the end since the beginning. We have watched the land taken by the sea and known that what is firm underfoot today might be gone tomorrow. Children used to learn about Noah's Ark and the Flood in Sunday school. They were taught the story so that they might pay attention to what is here now, and so seek to hold on to it. If only.

Cantre'r Gwaelod means 'the lowland hundred'. It is the stuff of legend in Cardigan Bay, in west Wales, where folklorists tell of a long-lost land of plenty, a veritable Eden of rich farmland protected from the sea by dykes and sluices carefully tended. Easy living it provided for long years in the first half of the first millennium AD – until complacency had some fool neglect the gates and sluices that were his responsibility. During a single awful night the land was flooded by the incoming tide, gone for good. A harder life had to be sought elsewhere.

A landscape of loss and lost places, then – ghost towns and villages that remind us that what we have now is only briefly ours. We know they're out there – beneath bumps in the grass, beneath the tide. What happened to theirs might happen to ours. It is from beneath the sea that we hear, on quiet nights, the sunken church bells tolling for us.

27

MOUNT COTTAGE, DORSET, AND A LIVING GHOST

'To die will be an awfully big adventure.'

J. M. BARRIE, *PETER AND WENDY*

'I heard the church bells hollowing out the sky,
Deep beyond deep, like never ending stars . . .'

JOHN BETJEMAN, *SUMMONED BY BELLS*

LATE IN THE GAME OF WRITING THIS BOOK, I SPOKE TO A WOMAN WHO SAID IT
sometimes seemed her family had lived in nothing but haunted
houses. Sasha Markov's roots are in England, but she lives now
near Los Angeles, California. She writes too. Screenplays and
such. It runs in her family. Her maternal grandmother was Alice
Mary Dilke (1919–2016) who, in the golden days of Hollywood,
was a reader in search of books and scripts that might be the stuff
of movies. She worked for Victor Gollancz, then Warner Brothers.
Doctor No was a novel she read and thought would make a good

film, though Warner thought not. Later she noticed a play by Frederick Knott and, in time, *Dial M for Murder* was directed for the big screen by Alfred Hitchcock.

Alice was, said Sasha, a reader of note, and also a thinker, always looking for new ideas. She married a fellow writer, Christopher Dilke, scion of a literary family. During the Second World War Christopher served in the Royal Artillery, rising to the rank of lieutenant colonel. After the war he worked for the BBC. He wrote books, and for film and radio. Alice and Christopher had four children together, two pairs born either side of the war – Caroline and Annabel before, Fisher and Lucy after.

From around 1941 until 1945, Alice rented Mount Cottage, in Dorset, with Caroline and Annabel, the oldest of their brood. Annabel would be Sasha's mother. Water came from a spring. There was no electricity, just candles and Tilley lanterns. Alice grew her own vegetables in the garden. Her family lived in nearby Hincknowle and her brothers – Giles, Sam and Henry – would come by and spend time with her. For all that it sounds idyllic, Alice remembered the house as a place of sadness. She was a young wife and mother. Christopher was away on active service, and she was raising their children alone.

After the war Alice and Christopher lived in a house in Campden Grove, in Kensington, London. Their two youngest were born there. In 1970 or thereabouts, Alice inherited some money and bought, for six thousand pounds, Vale House, near the village of Whitchurch Canonicorum, in Marshwood Vale in south-west Dorset. From around 1975 they lived there full time. Christopher died in 1987 and afterwards she lived alone, but for what Sasha described as 'a rotating cast of cats and dogs'.

Sasha loves the life she has in California with her husband and stepchildren. The air is clear, good for thinking and writing, she said. When we spoke, she had just bidden farewell to her cousin and her partner who'd been out for a visit. 'It's funny,' she said, 'because my cousin and I were talking about how, whenever we can't sleep, we go back in our minds to Granny's house in Marshwood Vale. We can remember every single detail of the house . . . the books on the coffee-table . . . everything.'

Sasha and I speculated then about whether it was a matter of the members of her family tending to form especially strong bonds to houses, imprinting on them and being imprinted by them in turn.

'Funnily enough,' she said, 'I was having a dream the other night . . .' She said that out in California her family have a close connection to a family of Navaho, and that for the Navaho, ghosts are regarded as a useful presence in life, an aid to understanding the world of the living. 'In the dream, I was having a conversation with a Navaho father, and he was saying what really has to happen in the US is that all the descendants of the settlers have to confront their ghosts. They say ghosts are a good way to talk about the past . . . The sins of the past . . . ancient trauma.'

Sasha said that for the Navaho, it's a lot easier to talk about trauma and emotion by way of ghosts.

Sometime in the early 1990s (Sasha was not sure precisely when) Alice was a guest at a party in a house in Dorset. At some point, she began overhearing a couple talking about how they had had to move out of their house on account of its being too haunted to live in any longer. The mention of ghosts was especially interesting to Alice, said Sasha, because by then she felt she had lived with one ghost after another.

'Everyone had always lived in haunted houses,' said Sasha. 'There were always ghost stories . . . always. I don't know if it surprised Granny . . . or any of us. There were just always ghosts . . . like members of the family. They just . . . kind of . . . These stories were not told in any shocking way.'

Alice had joined the couple's conversation then, asking which house in Dorset they were talking about. She had lived in the county a long time, knew it well and was curious to hear precise details. It was Mount Cottage, they said. Alice told them she had lived in that very house during the war.

'I don't remember any ghosts there,' said Alice.

'Well, there's a ghost now,' they said.

Alice asked the couple exactly what they had seen in Mount Cottage, the nature of the haunting so upsetting and frightening that they had had to move out. More than once, they said, they had seen a woman weeping beside the fire. The apparition, clear to see, was a young woman with brown hair. They saw her all the time. At first it sounded to Alice like standard ghost fare – the spirit of a young woman, sadness, so far so typical. There was more, though, said the couple. What was disturbing was something the ghost did every time. She would be weeping and then she would turn and bang her head against the wall by the fireplace, long enough and hard enough that blood would flow. Every night they would have to watch the weeping woman bang her head until blood spread across the wall, staining it like a red rose blooming. They could not take it any more so they had sold up.

Although not haunted, her years in Mount Cottage had given Alice different memories, a haunting of another sort. Back in the war there had come a day when Christopher had sent word asking her to go through his papers in search of something he needed urgently. Alice obliged, but in so doing ran across correspondence between her husband and another woman. It was plain to her that they had been having an affair. Already vulnerable on account of her isolation and loneliness, the news of the betrayal had been too much. She it was, then, who had stood by the fireplace and banged and banged and banged her head until she bled.

Alice was a ghost – *the* ghost. Still alive in the world and living elsewhere, she was haunting her old home. The couple had had no way of knowing what emotions she had experienced there. Only Alice knew that while living in Mount Cottage, a young wife and mother of two infants, she had learned of the infidelity of her absentee husband. Her pain had been close to unbearable. Sasha said depression was familiar to several members of her family, though seldom talked about: perhaps Alice had also been suffering anyway, post-partum. She and she alone remembered how she had banged her head against the wall until blood came. Although she had left Mount Cottage to live elsewhere – for the rest of her life and marriage, for the births of two more children – had she inadvertently left behind the essence of bitter betrayal and heartbreak?

Are we haunted not just by the dead, but by the living? Are we haunted by ourselves?

Sasha's father died when she was two years old. He was the Bulgarian dissident and writer Georgi Markov, murdered by the KGB in 1978 with a pellet of poison injected into his leg via the point of an umbrella; an infamous news story. Granny Alice had helped in the aftermath, helping Annabel raise her daughter. Sasha's cousin – she of the visit to California – lost her mother when she was young, and Alice helped with her upbringing too. She was a wonderful grandmother to all of her grandchildren, said Sasha, attentive and loving.

'She told us about a world of magic and understanding,' said Sasha. 'My grandfather died in 1987, and although she lived alone, Granny had tons of visitors. She was a wonderful, complex, deeply intelligent woman, a very deep emotional core . . . an emotional projection that was so strong. She was a formidable character and she loved us. She's kind of the reason we're all OK.'

Sasha had more memories of ghosts in the Markov family. Alice was often the source – reporting visitations in one house or another, in one case a ghost that visited her every night. It wasn't about being frightened or scared off. It was about an accepted presence.

After the house party in Dorset, Alice spoke about how she was haunting her wartime home. Word reached her younger children, Fisher and Lucy. It turns out they had their own memory of a childhood encounter, one they had not previously mentioned, in the Kensington house, in London, with an apparition dressed in black, sitting on the landing by the stairs. They had known it was their mother, their living and breathing mother, yet not her.

Sasha had said Alice was possessed of 'an emotional projection that was so strong', and I wonder if some sort of truth, or instinctive understanding at least, lies at the heart of it all. Was Alice unconsciously expressing her emotions in a form that adhered, like the scent of expensive perfume, to places it touched?

I asked Sasha if Alice ever made it to Movieland, the place where imaginings are made real, the industry to which she had meaningfully contributed.

'Alice never got to Hollywood,' she said. 'Or even to America. It was a big dream of hers to go. She was just a generation or two too early in terms of being a transatlantic creative career woman.'

Until I heard Alice's story, I thought all ghost stories were the same – or nearly. I have read scores of them, hundreds maybe. In almost every case, the explanation for the phenomenon is said to lie in the past. Wrongs – inflicted and suffered then – remain as glitches in the matrix, if you will, available like clips to be downloaded and witnessed over and over by inhabitants of one present after another. Or else the victim of the hurt, an avatar at least, exists in limbo, reliving the wrong and insisting the latest living live through it too, or pay some part of the debt in the form of their own upset and fear.

The story of the wee black car on the Isle of Skye was different: less a lingering of the past, more a portent of a tragedy to come; a tragedy that did come. For most of the rest the ghosts, or the stories of ghosts, serve best to answer a timeless need, or to hold at bay a timeless dread, which is fear of death. Anything, it seems, is better than death – even a dreadful ghost that offers proof at least of the survival of something after the fact, an immortality of a sort.

In *A Natural History of Ghosts*, Roger Clarke describes a 'very specific set of circumstances' that made England 'extraordinarily haunted'. The Catholic Church, notes Clarke, effectively appropriated unto itself the ancient, pagan belief in ghosts and, in the pragmatic way of those bent on absorbing and repurposing stubborn notions, had it that such were glimpses of souls in Purgatory. In the later Middle Ages, another iteration of Catholic Christianity clamped down on the ghost business, declaring it a grievous wrong even to talk of them.

Then, said Clarke, came the Reformation and the acquisition of all manner of monasteries, convents and the like. Monastic communities were scattered and, it was said, returned in death to their old haunts to haunt them. Where before such entities might have been taken in hand by the Church, shepherded like errant sheep, instead they were left to roam. Rather than souls in Purgatory, the Reformed Church taught that any so-called ghosts were evil spirits summoned by the Devil. For want of a church prepared to take ownership of them, such ghosts were understood to have gone rogue. Those susceptible to the stories found their own ways to accommodate the ghosts, and so make sense of them.

Clarke wrote: 'Belief in ghosts never became respectable, but its popularity was bolstered further by a new form of Christianity that also tacitly espoused a belief in spirits and ghosts – Methodism.'

It was by such a sequence of events that in Protestant Britain ghosts went through a metamorphosis. Rather than souls that might, in due course, find salvation and be gone, ghosts were seen as having nowhere to go, no hope of release, and persisted as the worst of sitting tenants. It was in this way that the foundations were laid for the plethora of haunted places, one church building and stately home after another with resident apparitions both religious and secular. Hence Iona Abbey and Windsor Castle, Glamis Castle and Raynham Hall, St Peter ad Vincula and Borley Rectory – and those are even before the more modern consideration of the need to make money from grandeur fallen on hard times.

The list is close to endless: add, for instance, Blickling Hall, in Norfolk, the childhood home of Anne Boleyn, that most peripatetic of phantoms. Her apparition appears there too, especially on

19 May, the anniversary of her execution in the Tower of London. Newton House, in Carmarthenshire, is haunted by, among others, the ghost of Lady Elinor Cavendish. Ordered to marry a man she did not love, Elinor ran home. Her husband-to-be came in hot pursuit and strangled her there.

During the 1980s a TV crew came looking for her too, and a camera operator swore blind he felt hands tight around his throat. Long before Bram Stoker made Whitby Abbey, in North Yorkshire, central to the tales of his Dracula, there were reports of the ghost of a novice nun who broke her vows and was bricked alive inside a wall for her pains; or Brighton's Preston Manor, another contender for Britain's most haunted, where there are more stories of hauntings than anyone could readily count. In 1896 a seance made contact with another nun, this one complaining of having been wrongly excommunicated and later buried in unconsecrated ground. A year later, work was carried out to improve the drains and the bones of a middle-aged woman were found in the process. After appropriate reburial in hallowed earth, all sightings of the White Lady promptly ceased.

On and on it goes, a litany of spirits, until Britain feels like a landscape smothered and muffled by a fog of the departed. In Bruce Robinson's *Withnail and I*, Uncle Monty observes that England is 'a kingdom of rains, where royalty comes in gangs'. Perhaps it is also a kingdom of the landlocked dead, where souls in limbo are accreting without hope of release.

Maybe if you bring ghosts, you find ghosts.

I see my dad when I'm not looking for him. He's in every mirror, looking back at me through my eyes. I have noticed, too, some of

Home to dreams and nightmares . . . Whitby Abbey

his mannerisms and foibles taking up residence. As I get older, I hear his turn of phrase sometimes when I speak. When there's a pause in a conversation, often with my kids, I am apt to say something like 'Aye, well, there you go.' It fills the space and nothing more. Dad did it. I never used to, but since he's been gone it's another of his quirks that's somehow become mine, as though it left him, like a poltergeist, and found a host in me.

There is more, always of an inconsequential nature: he used to ask for his tea and coffee in 'a thin mug'. He meant that he wanted to drink from something like fine china, rather than anything with a thick rim. Back when I still lived at home, in my teens, it used to annoy me as plain pernickety. Now I look for a thin mug too. I do not remember when that started. In one of his drawers my mum found recently bought and unworn socks of the sort he favoured, with toes and heels of a different colour from the main body. I find a strange comfort in wearing them, trying to fill his shoes, as it were.

It seems to me most obvious that belief in ghosts comes ultimately from fear of death. Anything being better than dying, might we cling instead to the idea of disembodied spirits in hopes that something of us survives, somehow? An eternity as a ghost, disturbing the living, might be a suboptimal life-after-death, but at least it's something, better than nothing, better than disappearing altogether.

And then I talked to Sasha Markov and heard Alice's story. What to make of the notion of being haunted not by the dead but by the living? Sasha had had more to say about the visit of her cousin and her partner to her home in California. Her guests were in bed, trying to sleep, when music they did not recognize and certainly had not selected started issuing from the speakers in their room. At first, they thought it was something technological, Wi-Fi speakers picking up and transmitting music from a neighbour's house perhaps. In any event they could not switch it off, but it was while they lay back on the bed that Sasha's cousin became aware of a figure in the room – a figure with flailing arms and legs, radiating anger. Her partner could see it, too, and they were terrified. It lasted for some moments, then disappeared.

'We talked and talked about it for days,' said Sasha. 'My cousin is a doctor, a scientist, but she was adamant about what they had seen.'

Early in our conversation Sasha said her family had lived in one haunted house after another, that ghosts were a given, and it seemed she meant it. Over the years, Sasha had learned to wonder if her family, given to dark moods, yet finding depression hard to talk about, had manifested ghosts as a means of externalizing their unhappiness and so making it easier to exorcize them.

She said her cousin had felt angry on arrival – and that after the incident in the guest room, her anger left when the flailing figure did likewise.

Elsewhere in this book, in the context of Culloden Moor, in the Scottish Highlands, I mentioned Archie Roy, he of the sense of something strange. Roy was professor of astronomy at Glasgow University, my own alma mater. He was one of Scotland's most distinguished astronomers, a Fellow of the Royal Astronomical Society and of the British Interplanetary Society. Such was Roy's expertise in the subject of orbital motion, NASA sought him out during the Apollo 11 programme and had him calculate the orbit of, among others, the capsule that carried Neil Armstrong and Buzz Aldrin in 1969. He was also a Fellow of the Society for Psychical Research and therefore known to some as 'Glasgow's ghostbuster'.

During the last ten years of his life, he dedicated a great deal of his energy to assessing the claims of mediums and clairvoyants. He felt that the mystery of how a medium might seem to glean, and ascertain, so much about a stranger was the sort of challenge best addressed by science. He also attended haunted houses – especially those apparently troubled by poltergeists. On my desk as I type these lines is Roy's *A Sense of Something Strange*, the book he wrote to catalogue the ghost stories that were to him most interesting. It was a Christmas present from my mum in 1990 and I have read it and reread it. 'Basically,' he wrote, 'the book contains just a few of the things that have over the years invoked within me a sense of something strange . . .'

On the subject of the paranormal, of ghosts, hauntings and the like, I am agnostic, by which I mean I cannot come down on either side of the debate. But I read about the wee black car on the Isle of Skye and I listen to Sasha Markov tell me about her grandmother Alice, and what I feel most strongly is that sense of something strange.

I read Archie Roy's words and am aware at all times that he was a man of science, who taught, for instance, spherical trigonometry to first-year university students. In an obituary in the *Guardian* newspaper of 3 February 2013, science writer Robin McKie described how Roy, who was ambidextrous, would stand in front of a blackboard and say, 'Let us take two intersecting circles,' and then, with a piece of chalk in each hand, proceed to draw two perfect overlapping circles. He was an ordered man and his was an ordered mind.

I read stories about ghosts and hauntings, and whenever I am tempted to dismiss testimony out of hand I think about death and recall what Roy, a man of science, said about life after death: 'If I die and I find out I have not survived, I will be very surprised.'

More than by ghosts, I am haunted by what death inspires in the minds of those left behind.

When J. M. Barrie's brother died, the thought of the lost boy took residence in his imagination. Peter Pan comes close to death near the end of the book. Exhausted by his fight with Captain Hook, he is with Wendy on a rock in the lagoon. The tide is rising and, with the last of his strength, he fixes her to the tail of a kite that can carry only one of them to safety. He is alone then, too tired to swim, far less to fly.

'To die will be an awfully big adventure,' he thinks.

Peter does not die, of course. Captured by the promise of eternal childhood, he does not even live, not truly. When Peter and Wendy meet for the first time and he tells her he has no mother, she is upset and reaches out to hug him.

'You mustn't touch me,' says Peter.

'Why?' asks Wendy.

'No one must ever touch me,' he says.

And therein lies the tragedy of Peter Pan, the little boy who never grows up and so never touches – or is touched in return – by any of what makes life worth living, which is growing alongside another and facing all that that means, old age and death included.

'The pain I feel now is the happiness I had before,' wrote C. S. Lewis as he grieved the loss of his wife, the poet Joy Gresham. 'That's the deal.'

28

FATHERS AND SONS

'Somebody said "True love is like ghosts, which everyone talks
about, and few have seen. I've seen both, and I don't know how
to tell you which is worse" . . . Well, I don't suppose you have to
believe in ghosts to know that we are all haunted, all of us, by
things we can see and feel and guess at, and many more things
that we can't.'

BETH GUTCHEON, *MORE THAN YOU KNOW*

ANOTHER DEAR FRIEND OF MINE TOLD ME WHAT HAPPENED WHEN HIS DAD DIED.
He was at his bedside in hospital. His dad was ill and failing.
Other members of the family had been coming and going, taking
turns to sit, awaiting the inevitable. As it happened, my friend
was alone with his dad when he gave up the ghost, as they say. My
friend is a straightforward and practical chap, not given to flights
of fancy. He has seen some stuff in his time, sobering stuff of the
sort that makes a person pay attention and also take life seriously.
He was a journalist in Rwanda in 1994 and witnessed some of the

genocide that took place there then at close quarters. More than once he has told me that, on account of what he saw, he no longer 'sweats the small stuff'.

When he told me what had happened when his dad died, he related the details in plain language, devoid of so much as a hint of the woo-woo. He said he was sitting by his dad's bed, and it started shaking – not gently vibrating but shaking as though an earthquake was taking place. He said his dad's bed shook and that he saw something like heat haze or perhaps smoke rise from his dad's body and disappear. The shaking stopped. Soon afterwards, a nurse came into the room, one of those who had been caring for my friend's dad. She confirmed he was dead, and my friend told her what had just happened. She said she was not surprised and that it was not the first time she had had such an experience described to her by those who had watched a loved one die.

Sadness gets a bad rap. Western society sells us happiness like it's the only state of being worth knowing. Happy feels good – in the same way that the effect of a few glasses of wine feels good. But it's hardly the best idea to be permanently pickled. I always wonder about those souls who say they are happy all the time. A little bit of happy goes a long way, but there are other ways to feel that are worth respecting. Sometimes sadness is the source of great creativity, great insight. Into each life, some rain must fall, and all that. Sadness gets such a bad rap in the West that millions of people are positively encouraged to spend their lives rattling with anti-depressants. As Danny the drug dealer says in Bruce Robinson's *Withnail and I*, to Marwood, overdosed and begging for some Valium: 'You have done something to your brain. You have made it high. If I lay ten mills of diazepam on you, it will do something else to your brain. You will make it low. Why trust one drug and not the other?'

I need happy and I need sad, and I try to see both for what they are. Kipling was on the money too:

If you can meet with Triumph and Disaster
And treat those two imposters just the same . . .

Happy and sad. Grief is another fellow traveller worth listening to. Fear of death is hosted by the part of us that doesn't grow up, or doesn't want to. I remember being afraid to go to sleep for fear I might not wake. My sister's bedtime prayer was:

As I lay me down to sleep,
I pray the Lord my soul to keep.
If I should die before I wake,
I pray the Lord my soul to take.

It worried me, that possibility of dying before I wake. I remember going downstairs in tears to find my parents and have them reassure me it was safe to close my eyes. I was afraid of Mum and Dad dying too. As well as their own deaths, children fear the death of parents, and surely parents fear the death of children most of all.

Fear of death and thoughts of ghosts are as old as writing. In the four-thousand-year-old *Epic of Gilgamesh*, a literary ancestor of the Old Testament, the eponymous hero king is devastated by the loss of his friend, Enkidu, to the Netherworld of ghosts. Grief sets Gilgamesh on a quest, the first recorded quest in search of immortality.

Despite his best efforts, he fails, and we read: 'Life, which you look for, you will never find. For when gods created man, they let death be his share, and life withheld in their own hands.'

Gilgamesh begs the gods to give back his friend and eventually Ea the sun god is persuaded to release him. The ghost of Enkidu rises, 'like a wind', through a tiny hole in the earth, and he and Gilgamesh are reunited and joined in an embrace. Gilgamesh is desperate to hear what it is like to be dead, to be a ghost, and at first Enkidu is loath to tell him, knowing the miserable truth will bring his friend nothing but sadness. Eventually he answers Gilgamesh's questions, describing the drab lot of the various dead.

'Have you seen a brother crying among relatives who chose to ignore his prayers?' asks Gilgamesh.

'Oh, yes,' replies Enkidu. 'He brings bread to the hungry from the dumps of those who feed their dogs with food they keep from people, and he eats trash that no other man would want.'

That quest undertaken by Gilgamesh is the archetype of the hero's journey – repeated by every other, ever after, from Egyptian Horus to Disney's Pinocchio and beyond – into the void, into dark, into the belly of the beast in search of wisdom, in search of the father each man must endeavour to find and also to become. Imagine: for as long as there has been writing, our species has written stories about death and ghosts, so they must be among the things that matter most. From the minute we could fix stories for all time, what did we save? As soon as we could write, what did it seem to matter to remember above all? Death and ghosts.

The antidote to death, and the fear of death, was offered most explicitly by Christianity, still the biggest religion in the world. Before being born as a man, Jesus had first to empty himself of his divinity – a process called kenosis. By so doing he ensured he would experience, in every detail, what it was to be human and alive, and so mortal. Before conquering his own death, he first brought dead Lazarus back to life. After dying on the cross, he returned in a form his friends and loved ones did not recognize at first, like a ghost. Soon after that crucifixion, on the road to Emmaus, some miles from Jerusalem, two of the Apostles were joined by that Third Man:

> And it came to pass, that, while they communed together and reasoned, Jesus himself drew near, and went with them. But their eyes were holden that they should not know him.

Even gods die; even God dies, and it evidently wears on their divine minds, on His mind. As far as we can tell, death was on God's mind from the beginning. In Genesis He ordered Abraham to sacrifice Isaac, the son he had waited a hundred years to have, for whom Sarah, his wife, had waited ninety years. God stayed the blade before blood was spilled, but why would a loving God contemplate anything so cruel? Except that as early as the genesis of all, he was rehearsing, with the help of his creation, his own death, the death that would be made manifest in the crucifixion of his only begotten son who had been with him from the beginning.

Again and again, God and gods have watched the worst happen. Odin of the Vikings had to watch his favourite son die too. Beautiful, beloved Baldr, made immune from harm by the oath sworn by everything in the cosmos to leave him be, was felled by an arrow shaped from mistletoe, the only thing that had not been asked to do so. The One God watched his son die and so too did the All-Father. There were hopes Baldr might be returned from the Underworld, but they were in vain. Always we have known and told each other about the inevitability of death, its infinite reach, and still we invest hope in remaining behind in the world of the living in whatever form. Is every glimpse of grey in every castle vault, every chill felt in a bedroom in a stately home, every sense of something strange, the emotional response of the fearful child in every one of us, clutching at hope?

Back on Sandwood Bay, in Cape Wrath, the far north-west of Scotland, there is the thought of a father and son bound together in life and in death. Of all the stories, that one haunts me more determinedly than most. The last action taken by the pair, the wrapping around themselves of the rope, is pragmatic and profound. No matter what happens, let us remain together. We can know nothing else about their relationship, how they were together, how they talked to or thought of one another. All that remains is the determination to stay together before and after death, the bond between father and son made unbreakable by the rope still wound around them when their bodies were found on the sand. We should be haunted by that image because it is unforgettable.

In that house of my mother and father, Dad said he saw an angel watching over her. In *The Waste Land* Eliot wondered about the identity of that other, the unidentified companion walking always alongside. German theologian Werner Pelz wrote that it is Jesus Christ – there, just as he was on the road to Jerusalem, beside those two disciples, in the days after his crucifixion. At the very least it is a hopeful thought.

We are haunted. The oldest stories make plain we always have been.

Come clean, old haunter,
Is that you not there?

ACKNOWLEDGEMENTS

This is the second book I have written for Transworld since we all left behind the old world and entered the new. I wrote *The Story of the World in 100 Moments* when we still had one foot in the world that existed before the Covid debacle. *Hauntings* is therefore entirely a product of where we are now, and the principal ghost of the book is made of all I used to think I knew.

Another book I wrote for Transworld, a few years ago now, was *The Story of the British Isles in 100 Places*. It came from a place of love – love of locations I had visited around these islands during the better part of the two decades I spent happily making television documentaries about archaeology and history.

Although I didn't declare as much in the pages of that volume of gentle stories, I also wrote it as a eulogy for something I believed had died – which is to say the version of Britain into which I was born and in which I had lived most of my life. Nothing and nowhere lasts for ever and at some point, before I started writing *100 Places*, it had occurred to me that *my* personal Britain was no more. The geographical entity is still proud of the sea, of course, but the identity and personality I understood are gone. I am a stranger here.

ACKNOWLEDGEMENTS

My Dad died too, while this book was still in the planning stages. I hadn't originally intended for him to feature in it, but there he is, old haunter. What I am trying to say, however clumsily and incompletely, is that in the writing of this book I am especially indebted to two personalities – one a man and one a place. *Hauntings* is therefore a book about ghosts and ghostly places. But along the way I realized I am haunted most of all by the past . . . which is to say my own past . . . the one my father lived in as well as me.

I am of course also indebted as always to my editor, patient and lovely Susanna Wadeson, for sticking with me through thick and thin. I owe, as well, a huge debt to Hazel Orme, for her painstaking and insightful handling of the text, and to Katherine Cowdrey for her astute picture research. The whole team at Transworld are a wonder to me, and to Barbara Thompson, Richard Shailer, Phil Lord, Liane Payne, Cat Hillerton, Tom Hill, Eloise Austin and all . . . I can offer only my grateful thanks.

My agent, Eugenie Furniss, at 42 Management & Productions, has been a constant, attentive presence for more years than I can count now. During the years since the world of before and the world of now, Eugenie is one of very, very few who have stayed with me. Thanks again, Eug, for everything.

My final and most grateful thanks of all go, as always, to my wonderful wife, Trudi, and to our babies – now big, old babies, right enough. Trudi is everything and everything comes from Trudi.

Any mistakes in the text are mine alone.

Picture Acknowledgements

Pages x–xi   Map drawn by Liane Paine

Page 13   Cape Wrath: Patrick Ward/Popperfoto/Contributor

Page 24   John Mulvany's painting of The Battle of Aughrim: Art Collection 2/Alamy

Page 34   The Dambusters' Raid: Christopher Furlong/Staff/Getty Images

Page 46   The Old Vicarage, Grantchester: Pictorial Press Ltd/Alamy

Page 60   The Young witch Jennet Device: Chronicle/Alamy

Page 72   Brocken Spectre on Helm Crag: John Finney Photography

Page 81   The island of Iona, Vintage etching circa mid-nineteenth century: powerofforever

Page 91   Old engraved illustration of Windsor Castle: mikroman6

Page 105   Glamis Castle, vintage etching circa mid-nineteenth century: powerofforever

Page 118   The Brown Lady of Raynham Hall: © Tallandier/Bridgeman Images

PICTURE ACKNOWLEDGEMENTS

Page 131   Culloden battlefield, Inverness: Karl Normington/
Alamy

Page 143   The execution of Anne Boleyn, 1536: Print Col-
lector/Contributor/Getty Images

Page 152   Toys for Annie: Image courtesy of The Real Mary
King's Close

Page 169   Borley Rectory: Charles Walker Collection/Alamy

Page 180   Wistman's Wood, Dartmoor: Devon and Cornwall Photography

Page 193   Wayland's Smithy: Peter Genet/Alamy

Page 204   The Angel of Mons: Chronicle/Alamy

Page 214   Lud's Church: Tim Cann/Alamy

Page 230   The assassination of William Horsfall: Photo 12/
Contributor Universal Images Group

Page 243   Robert Miller Oliver: Author's private collection

Page 258   Ivelet Bridge over the River Swale on the Corpse
Way: Neil McAllister/Alamy

Page 273   Corryvreckan Whirlpool, Jura: Nigel Housden/
Alamy

Page 279   The Three Sisters of Glencoe: Iain Masterton/
Alamy

Page 295   Montacute's troops in the subterranean passage,
Mortimer's Hole, Nottingham Castle: Historical
Images Archive/Alamy

Page 307   50 Berkeley Square, London: Rik Hamilton/Alamy

Page 319   Dunwich All Saints Ruins, 1904: The History Col-
lection/Alamy

Page 330   Whitby Abbey c.1870: The Print Collector/Alamy

INDEX

'NO' in the index indicates Neil Oliver.

Abbot's Bromley 185
Aberdeen 128, 129
Aberdeen Town Council
    138–9
Achallader Castle 282
Affleck, Isobel 70
Affleck, Tom 70–1
Age of Reason 236
agriculture 117
Akhenaten (Amenhotep IV),
    Pharaoh 186
Albert, Prince, Duke of
    York 106
Alexander, Michael 84
Allingham, William 181
Allison, Hugh G. 137
American Civil War 114
Amgueddfa Cymru
    (Museum Wales) 259
Amyatt, John 154
ancestor lights 222
Angel Row 293
angels
    American soldiers drowning 208
    George Washington's actions 207

Angels of Mons 200, 203, 205–6,
    207–8, 209, 212
Angles 193
Anglo-Saxons 79
Annandale Herald 162
Anne of Cleves 92
Anne of Denmark 57, 58
Anne, Queen 94, 131
Antarctic Reserve 247
Antarctica 247
Antrim 80–1
Aonach Dubh 279
Arawn 296
Archer, Lord Jeffrey and Mary 50
Argyll 81
Arrouaisian priests 4–5
Arthur, King 178
Arthur's Seat 153
artificial intelligence (AI) 227, 237
Ashopton 33–4
    Methodist chapel 35
Ashton, Professor John,
    CBE 253
Asser, Bishop of Sherborne 292
Aten 186

Auden, W. H. 63–4
Aughrim 18–30
  Battle of 18, 19, 23–7, 29
  Bloody Hollow 27–8
  Eachdhroim an áir 26
  hauntings of soldiers 27
  Kilcommedan hill 23
  killing field 27
  officer's greyhound 26, 28
  a Potter's Field (Dunton) 27
  Urachree morass 23
Aughrim Castle 23, 24, 26–7
Aughrim Fort 30
Augustinian nuns 288
Augustus, William, Duke of
      Cumberland 133–4
Ayrshire Yeomanry 242

Baldr 340–1
Balguy, Henry 34
Bamford 35
Baring-Gould, Sabine 183
Barker, Adam 258
Barker, Ann 258
Barnhill 272–4
Barrie, David 302, 303, 314
Barrie, J. M. 301, 302–4, 313–14
  Peter and Wendy 304, 321
  Peter Pan character 304, 334–5
Barrie, Jane Ann 303
Bath Society paper 205
bearded men 13–14
Beastmarket Hill 293
Beaulieu 115
Bégouën, Count 298
Beinn Fhada (long mountain) 279, 280

Bell Inn 293
Belloc, Hilaire 197–8, 198–9, 263
Ben MacDhui 66, 75, 76
  dome-shaped summit 66, 68, 71–2
  hauntings, Am Fear Liath Mòr
    67–70, 71–2
  shadows 71–2
  terrain 66
  unease and foreboding 67–9
Ben Nevis 66
Beowulf 194
Betjeman, John 321
Bidean nam Bian 280
Big Grey Man 67–70, 71–2
Billy Buck (Poole) 179
Birka Girl 64, 162
Björkö 64
Blickling Hall 328–9
Bloomsbury Group 45
Blount, Bessie 92
Blount, Sir John 92
Boleyn, Anne 92, 93, 94, 95–6, 141
  burial in Tower of London 149–50
  elusive truths about 141–2
  execution 145
  maligned 141
  marriage to Henry VIII 142
  miscarriage 145
  six fingers 141–2
Boleyn, George 95
Boleyn, Mary 92
bones
  of dead men 18–19
  for fertiliser 18, 19
  of Jacobites 26
  moss on 18, 19, 26

INDEX

Bonita's shop 208
Bonnie Prince Charlie 1, 130–1, 131–3
Book, Battle of the 80
Book of Kells 83
Booth, Frank 35
Borley Rectory 168–74, 328
  fire 168
  ghost hunters and sightseers 173
  hauntings 168–9, 170–1, 171–2
    ghostly nun 169–70
    human skull 170
    lightning strike 175
  paranormal activity 172
  priests 169–72
  public attention 173
  seance 175
Boswell, James 85–6
Bothvild 195
Bowes-Lyon, Claude, 13th Earl of
    Kinghorne 106, 108, 111
Bowes-Lyon, Elizabeth 106–7
Bowes-Lyon, Mary Eleanor 106
'The Bowmen' 202–4
  ghostly archers 203
  ghostly horsemen 204
The Bowmen and Other Legends of the
    War 203
Boyd, William, 4th Earl of
    Kilmarnock 149
Boyne, Battle of the 22, 29
'A Boy's Song' (Hogg) 100–1
Brando, Marlon 209
Brandreth, Jeremiah 231, 232, 233
British Army 202
British Expeditionary Force (BEF)
  202, 205

British Women's Temperance
    Society 305
Broad Marsh 300
Broadmarsh Shopping Centre 300
Brocken Spectre (Brocken Mountain)
  71, 74
Brocklehurst, Sir Philip 217, 219
Brontë, Charlotte 234
Brontë, Reverend Patrick 224–5, 226
Brooke, Rupert 44–5, 51
  Byron's heir 50
  death 44, 48, 50–1, 211
  education 45
  First World War 47
  illness 44, 45
  innocence 47, 48–9
  love life 45–6
  obituary 48
  poetry 45
  premonition 211
  travels 46
  works
    'Fragment' 49
    'Hauntings' 51
    'Nineteen-Fourteen' 44, 47
    'The Old Vicarage, Grantchester'
    46, 47
Brooke, Ruth 44
Brooke, William 44, 45
Browne, Sir Matthew 116
Brownell, Sonia 275
Bruce, William Spiers 247
Buachaille Etive Mor
    (Big Shepherd) 278
Buckland Abbey 182
Bulcock, John and Jane 57, 59

Bull, Edith 170
Bull, Harry 170
Bull, Henry 169
    daughters 170
Bunker's hill 156
burial grounds
    green burial corridors 253
    Pendle Hill 53
    St Mahew's Chapel 262–3
    Wayland's Smithy 191–2
    see also cemeteries
burials
    polluting nature of 260
    rites 187
    Scottish tradition 262
Byron, Lord 50–1, 234, 236
    death 50
Byron's Pool 50
    Byron's ghost 50

Cabell of Buckfastleigh, Squire
        Richard 183
Cailleach Bheur (Hag of
        Winter) 273
Cairngorms 69, 70
Caledonian Canal 129
Callanish 267
Calton hill 156
Cambuskenneth 4–5
    hauntings 5–6
Cambuskenneth Abbey 4–5
Cameron, Donald 132, 136
Cameron of Lochiel 101
Campbell-Bannerman, Henry 305
Campbell of Ardkinglass, Sir
        Colin 283

Campbell of Glenlyon, Captain
        Robert 284, 285–6
Campbell of Glenorchy, John, 1st Earl
        of Breadalbane 282
Campbell of Skipnish,
        Archibald 104
Canadian Rockies 68
Canning, George 309
Canterbury Cathedral 198
Cantre'r Gwaelod 320
Cape Wrath 10–12
    lighthouse 12
Cappadocia 296
Carlisle, Bishop of 256
Carlyle, Thomas 233–4, 237
Carmelite monks 293
Carmichael, Alexander 286–7
Carmina Gadelica (Carmichael)
        286–7
Carnock Burn 220
Cartwright, William 226
The Castle of Otranto (Walpole)
        120, 236
Castle Rock 152, 153, 156
Castlehill 58, 102, 103–4, 104
Catherine of Aragon 91–2, 142
    death and funeral 144–5
    marriage to Henry VIII 143
    missive 144
    pregnancies 144
    rejected Anne as Queen 145
Catholic Church 327
cattle thieves 279, 280
Cave of Les Trois Frères
        297, 298
Cavendish, Lady Elinor 329

INDEX

caves 292, 297–8, 299–300
  see also Tigguo Cobauc
Celts 77–8, 79
  Celtic roots 77–8
cemeteries 253–4, 261
  Flanders 51
  graves 260–1
  Reilig Òdhrain 87
  resources 260
  Scarp burial ground 261
  see also burial grounds
Cenotaph 51
Chaillot, nuns of 288
Changi Air Base 31
Channel Tunnel 129
Chapel of St Edmund 288
Chapuys, Eustace 142
Charles I, King 20, 93, 94, 98–9, 280
Charles II, King 21, 116, 280
Charteris, Brigadier General
    John 200
Chattox Clan 56
  Anne Chattox 56, 57
Cherry-Garrard, Apsley 246
Chesterton, G. K. 151
Christianity 40, 80, 83, 339
Church of England 92
Church of St Mary the Virgin,
    Muker 257
Church of the English Benedictines
    288
Church of the Twelve Apostles 89
Churchill, Winston 45, 48
Clachaig Inn 286
Clan Cameron 101
Clare, Richard de 'Strongbow' 20

Clarke, Roger 327–8
Clavert, Robert 224
Clearances 10
  Picts 10
Clitheroe Castle 53
Clunies-Ross family 164
  George 164, 166
  John 164, 166
  John-Cecil 164
  John George 166
  John Sydney 166
Cnoc an Airm 263
Cnoc nan Aingeal (Hill of the
    Angels) 83
Coast Australia 163–4, 165–7
Cocos Islands 163–4
  inhabitants 165
  Malay workers 164, 165
  ownership of 164
coffin roads 178, 182, 253–65
  brides' winding sheets 264–5
  crossroads 257
  east-to-west direction 262
  hauntings 264
    black dog 257
  loss of 256
  pall bearers 254–5, 263
  restless spirits 257
  Scottish 262–4
  Stoneymollan 262–3
  strangeness 264
coffin stones 257
coffins 255, 257
  cairns 263
  chemicals used in 260
  flat stones 255, 263

coffins – *cont.*
  live occupant 264
  valuable resources for 260
  wicker 257
Coire Gabhail 280
Collie, Professor J. Norman 67–8,
    69, 71
Commonwealth 20–1, 106, 280
Commonwealth War Graves
    Commission 261
Connolly, Cyril 275
Contact (Sagan) 40
Cooper & Co's Stores Ltd 242, 244
COP$_{27}$ 237–8
Corn Laws 231–2
corpse-candles 258–9, 259,
    260, 261
Corryvreckan 273, 274
Country Life 117, 123
Covid lockdowns 238
Cranmer, Thomas 142
Crean, Tom 245, 246, 249
Cromwell, Oliver 20, 20–1, 99, 106,
    227, 280
Cromwell, Thomas 142
croppers 228
Crowley, John 314
Cúl Dreimhne 80
Culloden 101
Culloden, Battle of 127, 128, 130
  destruction of Gaelic Highlands 134
  hauntings 138–9
    dead Highlander 134–5
    gunfire and cries of men 135
    Jacobites 135
    Skree of Culloden 136, 137–8

tall man 134
Two Sights visions 127, 128,
    138–9
washerwoman (*bean
    nighe*) 136
  political motivations 130
  slaughter and atrocities 133–4
Culloden House 136
Culloden Tales (Allison) 137
Culpeper, Thomas 95
Curzon, Elizabeth 309
Custom House Quay 18
Cymru 79

Daedalus 195–6
Daemonologie (James I) 58–9
Dál Riata 81
Dalrymple, Sir John 282, 283–4,
    285, 286
The Dam Busters 32
Dambusters' Raid 36–8, 40
The Dark Is Rising (Cooper) 179
dark, lure of 292–3, 298
Dartmoor 181, 182–3
Darwin, Charles 165
David I, King 4
Davis, Andrew Jackson
    'Poughkeepsie Seer' 112, 114
'Day Is Dying In the West'
    (Lathbury) 35
death, fear of 338, 339
Demdike family 57
Derbforgaill 20
Derby, Bishop of 35
Dereham, Francis 95
Derinkuyu 296

Derwent Hall 34–5
Derwent Reservoir 33, 34–5, 38–9
  flood 33, 35, 36
Derwent Valley 39
Derwent Valley Water Board 35
Desmet, Mattias 207
Devereux, Paul 221, 222
Devereux, Robert 149
Device, Alizon 55–6, 57, 59
Device, Elizabeth 56, 57, 59, 62
  Ball (pet dog) 62
Device, James 56, 57, 59
Device, Jennet 59, 61, 62
Device, John 56
Devil's Pulpit 220
Dial M for Murder 322
Dian Cecht 220
Diarmait 80
Dickens, Charles 310, 313
DiFrancesco, Ron 249–50
  escape 250–1
  fireman 251
  hospital 251
  survivors' guilt 251
  voice 250
Dilke, Alice Mary 321, 322, 323,
    324–5
  ghost of 325, 326
Dilke, Annabel 322, 325–7
Dilke, Caroline 322
Dilke, Christopher 322, 324–5
  infidelity 325
Dilke, Fisher 322, 326
Dilke, Lucy 322, 326
Diodorus of Sicily 267
divine intervention 207, 210

Doctor No 321
Doctrine of Signatures (*signa naturae*) 28–9
Doidge, William 208, 209
  hoax 209
Donne, John 89, 287
Doug (WWI veteran) 208
Douglas, Archibald, Earl of
    Angus 103
Douglas, Lady Janet, Countess of
    Glamis 102, 103–4
  witchcraft 104, 110
Doyle, Arthur Conan 182–3
Drake, Sir Francis 182
Druids 183–4, 216, 220
Drummond, Thomas 284
Drumossie Moor 129–30, 136
Dudley, Guildford 146
Dudley, John, Duke of
    Northumberland 149
Duguay-Tourin 44
Dumb Steeple 225
Dumfries 301–2
Dumfries Academy 301, 302
Dun Bull Hotel 255
Duncan, Geillis 57
Dundas, Captain J. D. 148
Dunton, John 27
Dunwich 318–19

Eadred, King 194
earth lights 221–2
Eastern Daily Press 204–5
Edersee dam 37, 38
Edge, William 116
Edgewood 314

INDEX

Edinburgh 151–2
  burials 156–7
  City Chambers of Edinburgh
    Council 156
  class system 155
  crag and tail 153
  expansion of 154
  hills 156
  Nor Loch 154
  plague 156
  Princes Street Gardens 154
  process of confinement 153
  Royal Exchange 155, 156
  Royal Mile 58, 102, 151, 153, 155
  skyscrapers 153–4
  Underground City 152
Edinburgh Castle 102
Edward II, King 294
Edward III, King 90, 294
Elephant Island 245, 246, 249
Eliot, T. S. 249, 252, 341–2
Elizabeth I, Queen 90, 98, 228, 280
Elizabeth II, Queen 89, 91
Elma (NO's aunt) see Oliver, Neil:
  aunt (Elma)
Elphinstone, Arthur, 6th Lord
  Balmerino 149
embalming process, toxic
  chemicals 260
emotional residue 40, 111
enclosures, land 54
Endeavour 187
Endurance 245, 247
'Endymion' 249
Enkidu 338, 339
Enlène Cave 297, 298

The Face of Britain 99–100
The Fairies (Allingham) 181
Faro 222–3
Faulkner, William 43, 176
Felton, John 99
50 Berkeley Square 306–9, 311, 313, 314
  ghosts 312, 313
Finnian 80
Finnich Glen 220
First World War 43, 47–9, 115, 242
  cemeteries 51
  sadness of 51
fishermen 16–17
  legends of wreckers 16–17
  plundering and murdering
    of 16–17
  respect for the sea 16
Fitzroy, Henry 92
Fitzsimmons, Mr 288–9
Flanders, cemetery 51
Fletcher, Robin 272
Flodden, Battle of 142, 153
Flood, the 320
Forbes, Duncan 135–6
Forbes, John 103
Forest of Bowland 52, 55
Forest of Windsor 96
Forster, E. M. 45
Fort William 281, 283
'45 rebellion 131–3, 136
Fourth Industrial Revolution 300
fourth man phenomenon 249
Fox, George 52, 53
Fox, Kate 114
Fox, Leah 114
Fox, Maggie 113–14

INDEX

Foyster, Adelaide 171
Foyster, Marianne 171, 172, 175
Foyster, Reverend Lionel Algernon
        171, 172, 173
'Fragment' (Brooke) 49
frame breaking 234
*Frankenstein; or, The Modern
        Prometheus* (Shelley) 235–6
Fraser, James George 184
Fraser, Simon, Lord Lovat 149
free floating anxiety 207
French Revolution 288–9
Frere, Richard 68, 69, 71
Fussell, Paul 50

Gallipoli 49
Gallows Hill 61
Gawain 215, 223
Gazetteer of Scottish Ghosts 12
Gearr Aonach (short ridge) 279, 280
Geiger, John 249, 251
Geoffrey of Monmouth 219
George III, King 90, 94, 97–8
George IV, King 90, 289
Ghost Houses (Harper) 311–12
The Ghost in the Machine (Koestler) 238
'The Ghost in the Machine' (Tandy
        and Lawrence) 73, 74–5
ghost stories 1–3, 252
ghosts 1
    fear of 338
    50 Berkeley Square 312, 313
    Glencoe 286–7
    Greeks 186
    Henry VIII 94–5
    Iona 80, 88

Lud's Church 223
Mount Cottage 324, 325
Navaho 323
origins of 185–6
reasons to believe in 162
returning home 65
Romans 180
stately homes 115
Gibo, Aiko 160
Gibson, Guy 37, 38
Gilgamesh 338–9
Ginkel, Godert de 23–6
giolla coluim mac an ollaimh 277
Glamis Castle 328
    appearance 105
    hauntings
        African servant boy 107
        armoured knight 107
        Black servant 107
        Hangman's Chamber 107
        Lady Janet Douglas 110
        Lord of Light 110
        Monster of Glamis 108–9
        old lady 107
        Tongueless Woman 109
    history 105–6
    mistreatment and restorations 106
    sadness 108, 110
    secret 106, 108, 109, 111
    Sir Walter Scott 110–11
Glamis, Lord Thomas 108, 109–10
Gleann-na-Fola 25
Glencoe 290
    ghosts and apparitions 286–7
    green life 278–9
    meaning of 277–8

355

INDEX

Glencoe – *cont.*
  mountains of 279–80
  rock of 278
Glencoe Massacre 284–6
Glorious Revolution (1688) 19, 131
God 340
The Golden Bough (Fraser) 184
Gollancz, Victor 321
Good Queen Bess *see* Elizabeth I,
     Queen
Gordon, Seton 11
Gothic fiction 236
Gradbach Hill 213
Graham, John, Viscount Dundee 281
grand houses 113
graves 226, 260
  items of value 260
  respect for the dead 260–1
  rites 254
Great Expectations (Dickens) 310, 313
Great Glen 129
Greek War of Independence 50
Greeks, ghosts 186
Green, Andrew 14–15
green burial corridors 253
Green Chapel 215
Green Knight 215
green, symbolism of 216
Gregory VIII, Pope 294
Gresham, Joy 335
Grey, Alice 57, 59
Grey, Henry, 1st Duke of Suffolk 149
Grey, Lady Jane 146, 149
Greyfriars 300
Grice, Katherine 227
Grime's Graves 190, 297

Grinton Church 258
Grote Mandrenke (Great Drowning
    of Men) 317, 318
gui 187
Gunn, Alexander 11
Gutcheon, Beth 336
gweilo 187
Gwydion 178

Hades 296
Hamilton of Finnart, Sir James 103
hanter 65
Haqel D'ma 27
Harper, Charles 311–12
Harris, Arthur 'Bomber' 38
Harris, Isle of 264
Hartshead 224, 226
A Hat Full of Sky (Pratchett) 194
hauntings 63–5
'Hauntings' (Brooke) 51
Hawaiki 188
Haweswater Reservoir 255–6
Hawkins 120–1
Hearn, Paddy 268
Hebrides 266–8
heimta 65
Henderson, Jan-Andrew 155
Henry II, King 20
Henry VIII, King 54, 94, 142,
    227, 293
  behaviour towards women 93
  ghost of 94–5
  head of the Church of England 144
  injury 95
  King's Great Matter 92
  mistresses 92

INDEX

wives
    Anne Boleyn 92, 93, 94, 95–6,
        141–2, 145, 149–50
    Anne of Cleves 92
    Catherine Howard 92, 93–4,
        94–5, 149
    Catherine of Aragon 91–2, 142,
        143, 144–5
    Catherine Parr 92
    Jane Seymour 92, 93, 94, 145
Hephaestus 190–1
Hera 190, 191
Heriot hill 156
Herla 178
Herne the Hunter 96–7,
    98, 178
Hewitt, Katherine 57, 59
Highlands 134
  peace 282
  roads 270
Hillaby, John 254–5
Hinduism 315–16
Historical Library 267
History of the Kings of Britain
    (Geoffrey of Monmouth) 219
Hitchcock, Alfred 322
Hogg, James 72, 100
Holland, Richard 258–9
Homage to Catalonia (Orwell) 274
Home Island 165
Homo sapiens 189
Hood Battalion 49
Hood, Robin 229
Horn Dance 179, 185
Horne, Richard 97
Horsfall, William 229

The Hound of the Baskervilles (Doyle)
    182–3
House of Hanover 131, 136
Housman, A. E. 211
Howard, Catherine 92, 93–4, 95, 149
Howden Reservoir 35, 36
Hunter, Will 299
Hyperboreans 267

Icarus 195
Imperial Trans-Antarctic Expedition
    245–7, 248–9
Io 188
Iona 79–80, 81–2, 85–6
  beauty of 82
  Camus Cúl an t-Saimh 82
  Carn ri Eirinn 82
  Christianity 83, 84
  ghosts 80, 88
  Lewisian Gneiss 82
  Martyrs' Bay 84–5, 87
  monastery and university 82–3, 85
  monks 79–80, 84, 87, 88
  myth 82
  pilgrims 82
  Port-a-Churaich 82
  Reilig Òdhrain (Relics of Oran)
    87, 261
  Sraid nam Marbh (Street of the
    Dead) 87–8, 261
  Torr an Aba (Hill of the Abbot)
    86–7
  Vikings 84
  works of art 83
Iona Abbey 328
'The Iona Boat Song' 87

Ireland 20
   Battle of Aughrim 18, 19, 23–7, 29
   Catholics and Protestants
      20–1, 22
   English rule 20–1
      rape of land 20–1
   Jacobite Parliament 22
   Norman conquest 20
   Williamite War 19–20, 21, 22, 29
'Is It OK to Be a Luddite?' 234
Isabella, Queen 294
Isabella (wife of Earl of Seaforth) 138
Iscariot, Judas 27
islands 272

Jacobites 22, 132, 133–4, 135, 281
James Caird 245
James, Clive 41–2
James, Henry 45, 315, 318–19
James I, King 54, 57, 58–9, 59, 228
James II, King 29, 131, 132, 153, 280,
      287–8
   coffin, desecration of 288–9
   corpse and body parts 288, 289
   death 288
   Williamite War 19, 21, 22, 29
James III (Old Pretender), King 5, 131,
      280–1, 282–3
James IV, King 5, 142
James V, King 102–3, 104, 106
James, Simon 77
Jenkinson, Robert 232
Jesuits of St Omer 288
Jesus 296–7, 339–40, 342
Johnson, Dr Samuel 85
Jura 272–6

Kali 315–16, 319–20
katabasis 296
Kaye, Tony 208–9, 209
Keats, John 249
Keeling, William 164
Kenneth I, King 4
kenosis 339–40
Kent, William 116, 119, 306–8
Kentfield, Captain 313
Keynes, Maynard 45
Kilcreggan 263
Killiecrankie 281
Kimbolton Castle 144
King Haakon Bay 246
King, Martin Luther Jr 315
King of the Wood 184
Kinghorne, Earl of 106
Kingsley, Charles 17
Kipling, Rudyard 196, 197,
     198, 338
Kirklees Priory 227
Knavesmire 59
Knott, Frederick 322
Koestler, Arthur 238
Kyle of Lochalsh 268, 271
Kyleakin 269
   ferry 270
   hauntings
      little black car 270–2

Ladybower Reservoir 35, 36, 255
   haunted place, reputation 33
   hauntings
      Dambusters 33
      Lancasters 33
Lairig Ghru 66, 69

INDEX

lakes
  Derwent Reservoir 33, 34–5, 36, 38–9
  Haweswater Reservoir 255–6
  Howden Reservoir 35, 36
  Ladybower Reservoir 33, 35, 36, 255
  Lower Black Moss Reservoir 54
Lancasters 38, 39
The Last Words of Distinguished Men and Women (Real and Traditional) (Marvin) 288
Lathbury, Mary A. 31, 35
Law, Abraham 55
Law, John 55–6, 59
Lawrence, T. E. 175
Lawrence, Tony R. 73
Lazarus 340
Lee of Calveron, William 228
St Leonard's hill 156
Leverthorpe, Joan 227
Lévi-Strauss, Claude 190
Lewis, C. S. 335
life and death 264–5
Lindisfarne 83
Lindisfarne Gospels 83
Lindsay clan 107–8
Lisbeg 30
Little, Big, (Crowley) 314
Little Guide on Devonshire (Baring-Gould) 183
Lively, Penelope 176, 179
The Living Mountain (Shepherd) 66
Livingston, Thomas 283
Llewelyn Davies family 303–4
Llud Llaw Ereint (Llud of the Silver Hand) 219–20

Loch Roag 267
Loch Sandwood 9
Lockerbie 162
Loftus, Colonel 120–1
Lollards 216, 216–19
Lollardy 217, 218
longing 65
Longleat 115
Louis XIV, King 23, 281
Lovett family 305
low-frequency sound waves 73–4
Lower Black Moss Reservoir 54
Lud-Auk, Alice 217–18, 219, 221, 223
Lud-Auk, Walter de 216, 217, 218, 219, 223
Lud, King 219–20
Ludchurch 217
Ludd, Ned 228, 229
Luddites
  awareness of change 229–31
  burials 227
  Byron speaks on behalf of 234
  craftsmen and skilled workers 228, 229–31
  deaths 227, 229
  deportations 231, 232
  evictions 232
  hangings 229, 231, 232
  high treason 232
  laughing at 227–8
  Pentrich Revolution 231–3
  protests and rebellion 229
  Rawfolds Mill confrontation 225–6
  torching of factory 231
  unskilled workers 231

Luddites – *cont.*
  vandalising machinery 228
  violence against 227, 229, 232–3
Ludlow, David 205
Ludlow, William George 204, 205
Lud's Church 213, 214–16
  dark reputation 221
  de Lud-Auk family 216–19, 223
  ghost lights 221, 222
  ghosts 223
  ill-will 216
  inspiration for *Sir Gawain and the
    Green Knight* 215–16
  stories 220
Luke's gospel, New Testament 252
Luss 263
Lutyens, Edwin 51
Lych Way 182, 255
Lyon-Bowes, Thomas 108–9
Lyon, John, 1st Thane of Glamis 106
Lyon, John, 6th Lord Glamis 104
Lyon, John, 9th Earl of Kinghorne 106
Lyon, Patrick, 1st Lord Glamis 106

Macdonald of Glengarry 282
Macdonalds of Glencoe 279, 283,
  284–5, 287
MacFarlane clan 263
Machen, Arthur 202–4, 206, 209, 212
MacIain Macdonald, Alasdair, 12th
  Chief of Glencoe 282–3, 284–5
MacIain Macdonald, Alasdair (son)
  284–5
Mackenzie, Francis Humberston 138
Mackenzie, George, 3rd Earl of
  Cromartie 149

Mackenzie, Kenneth 'Dark Kenneth'
  128–30
  Brahan Seer 129, 134, 135, 138, 264
  visions 129
MacLean, Elizabeth 305
MacLean, Reverend Ebenezer 305
MacLennan, Donald 261
MacLennan, Donald John 261
MacLeod, George 85
MacLeod, Kenneth 266
MacMurrough, Dermod 20
MacNeice, Louis 63–4
Maggs Bros Ltd 306, 313
Maggs, Uriah 306, 310
Malkin Tower 56, 57
Man Seeks God (Weiner) 40
Mantel, Hilary 142
Maoris 187–8
Mardale 255
Mardale Green 255, 256
Mardale Holy Trinity 255–6
Mare, Walter de la 309–10
Margaret of Denmark, Queen 5
mariners 12
Markov, Georgi 325
Markov, Sasha 321, 322, 323, 324,
  325–7, 331–2
  hauntings
    flailing figure 332
    haunted houses 332
    memories of ghosts 326
Marquesses Townshend 115
Marryat, Captain Frederick 121–3
Marryat, Florence 121–3
Marston Moor, Battle of 227
Martyrs' Bay 84–5, 87

INDEX

Marvin, Frederic Rowland 288
Mary II, Queen 19, 22, 131, 281
Mary King's Close 151–3, 155
  Chesney family 158
  hauntings 157–8
    Little Annie's doll 151–2, 160
    Major Thomas Weir 158, 159
  inhabitants buried alive 156
  sealed up 156
  tourist attraction 157
mass formation psychosis 207
Mayerling, Louis 173–4, 174–5
Mayfair magazine 301, 312–13
McCarthy, Tim 245, 246
McCombs, Phil 210
McGough, Roger 294
McKie, Robin 333–4
McNeish, Harry 'Chippy'
  245, 246
Mearns, William Hughes 163
mermaids 11
The Merry Wives of Windsor
  (Shakespeare) 96
Methodism 328
Millerism 114
mines 297
Minotaur 196, 296
Mochta 263
Möhne dam 37, 38
Mons, Battle of 202, 204, 207,
  211–12
  angels and phantom archers 203,
    205–6, 207–8, 209, 211
  shining hope 211
Montagu, Lord 115
Montair, Henry 217, 218–19

Montesquieu-Avantes 297
Moody Blues 261
More, St Thomas 149
Mormonism 114
Morris, Chris 209
Mortimer, Sir Roger 294
Mortimer's Hole 294
moss
  as cure 29
  on dead bodies 18, 19, 26
Moss Man 216
The Most Haunted House in England:
  Ten Years of Borley Rectory
  (Price) 171, 173
Motley, James 258–9
Moultree hill 156
Mount Collie 68
Mount Cottage 322, 324
  ghosts 324, 325
Mount Tamboro eruption 235
mountain panic 67
Muir, John 10
Mull, Isle of 80, 85
Munro, John Leslie 39
Murray, Lieutenant General Lord
  George 136–7
Myers, Thomas 309, 310–11, 312

Nanga Parbat 68
Napoleon 225, 306–8
National Service 31, 32
A Natural History of Ghosts
  (Clarke) 327
Navaho 323
  ghosts 323
Nechtan 178

INDEX

Neill, James Cameron
    (NO's grandfather) *see* Oliver,
    Neil: grandfather (James
    Cameron)
Nelson, Admiral Horatio 37
Neolithic farmers 191–2
Nevill, Lady Dorothy 309
Nevis Bank 268
New Model Army 21
New Scientist 255
New York World 114
Newton House 329
NHS (National Health Service) 238
Nidud of Sweden 191, 195
Nine Years War 281
Nineteen Eighty-Four (Orwell) 273, 275
"Nineteen-Fourteen" (Brooke) 44, 47
    'Peace' 47
Noah's Ark 320
Norman, Sir Montagu 175
nostalgia 65–6
Notes and Queries 147
Nottingham Castle 293
Nottingham, caves 292–3, 299–300
Nowell, Robert 56, 57
nuclear war 237
Nunsbrook 227
Nutter, Alice 57, 59–61

oak trees 183–4
Oceania House 164, 164–5, 174
    filming 165–6
    hauntings 165, 166, 167–8
Odin 178, 340
Odysseus 296
Oedipus 190

Ogilvie clan 107–8
Ogilvy, Margaret 302–3
Old Demdike 56, 57, 59
The Old Road (Belloc) 197–8, 198–9
Old Vicarage, Grantchester 45, 50
    Brooke's ghost 50
'The Old Vicarage, Grantchester'
    (Brooke) 46, 47
Oliver, Neil
    aunt (Elma) 161, 251–2
        'Sonny', invisible friend 161, 251–2
    children 101
    Coast Australia documentary
        163–4, 165–7
    daughter (Evie) 305–6
    DNA 100
    family homes 99–100, 176, 301–2,
        305, 313–14
    ghosts 305–6
    father 33, 244, 276
        army photo 31–2
        death 65, 176–7
        funeral 101
        hands 6–7
        Highlands trips 268–9
        illness 239–40
        Isle of Skye trips 269
        little people ghosts 306
        mannerisms and foibles 330–1
        memories of 65
        Norma's guardian angel 201, 212
        RAF uniform 244
        smoking 32
        thin mug 331
        travelling 99–100, 101, 268
        travelling salesman 268

INDEX

grandfather (James Cameron) 43, 125–6

grandfather (Robert Miller) 212
  appearance 244, 245
  army 242
  gold wristwatch 242, 244
  grocer's shop 242, 244
  shrapnel 244

grandmother (Peggy) 125, 126, 161, 162
  little James 161

great-aunts (Jessie and Madge) 125

hauntings 63–5, 248, 269

lost children, thoughts about 162

Maori documentary 187–8

mother (Norma) 33, 65, 101, 125, 126, 201
  ancestry 100, 101
  family 161
  group of boys vision 241
  macular degeneration 240–2
  red-brick house vision 241

nephew (Sonny) 160–1

Oceania House experience 166–8

sleep paralysis 167

uncle (Andy) 125

wife (Trudi) 101, 177

William Spiers Bruce documentary 247–8

Oliver, Norma Agnes Cameron (née Neill) see Oliver, Neil: mother (Norma)

Oliver, Robert Miller (NO's grandfather) 43, 212

Ollerenshaw, Olive 35

Operation Chastise 36–8

Orwell, Avril 272

Orwell, Eileen 272

Orwell, George 272–6

Orwell, Richard 272, 273–4

Owen, Wilfred 49–50

The Owl Service (Garner) 179

Palace of Holyroodhouse 62

Pankhurst, Emmeline 305

Paracelsus 29

Parr, Catherine 92

Pathology of Boredom 241

Pearless, Frank 172

Peel, Frank 225–6

Peggy (NO's grandmother) see Oliver, Neil: grandmother (Peggy)

Pelagic 247

Pelham, Elizabeth 119

Pelz, Werner 342

Pendle Hill 52–3
  Big End 53
  Devil's Apronful 53
  exploitation of people 55
  hauntings
    airmen 61
    dark dog 62
    witches 61
  Pendle Grit 54
  sadness 61–2
  stories 53
  views from 53–4
  witch story 54
  witchcraft 55–7, 59–60

Pendle Witch Trials 57

Pentrich Revolution 231–3

INDEX

Perceval, Spencer 234
Peter and Wendy (Barrie) 304, 321
Philippa of Hainault 90
Phrygian culture 296
The Pickwick Papers (Dickens) 313
Pictorial Guide to the Lakeland Fells
    (Wainwright) 255
Picts 10, 79
Pio, Padre 210
    miracles 210
    redirected bombers 210
Pitchfork Rebellion (1685) 149
places
    defence of 13–14
    emotional residue 40
Pliny the Younger 186
poison gas 206
Pole, Margaret 146–7
Polidori, John William 234–5, 236
Port Stanley 247, 248
Prebble, John 127, 138–9
Preston, Jennet 57, 59
Preston Manor 329
Prestonpans 132, 137
Price, Harry 171, 173
Proby, Mary 138
property
    defence of 13–14
    hauntings 13–14
protectionism 228
Provand, Captain Hubert C. 112, 117
Public Universal Friend 114
Puck (Kipling character) 196,
    197, 198
Puck of Pook's Hill (Kipling) 196
Pwyll, Prince of Dyfed 296

Pynchon, Thomas 234, 235, 236,
    237, 238

Quaker movement 53, 61
Quant, Mary 151–2

RAF 31–2
Ramsay, Andrew Lord
    Abbotshall 159
Ramshaw Rocks 213
Rannoch Moor 278
Rattan, Suruchi 162
Ravenser Odd (Hrafn's Eyr)
    316–17
Rawfolds Mill 225–6
Raynham Hall 115–17, 328
    architecture of 116
    hauntings
        Brown Lady 117–19, 119–23
Raynham, Lord 120
Redferne, Anne 57, 59
Reformed Church 328
Reilig Òdhrain (Relics of Oran) 87
Richard II, King 96
Richard the Lionheart 293–4
Ridgeway 193, 197, 198
*The Risings of the Luddites, Chartists
    and Pluggers* (Peel) 225–6
River Dart 184–5
Roaches, the 213
Robert II, King 106
Robinson, Bruce 329, 337
rock art 222
Romans 10, 84, 152
    ghosts 186
Rosneath churchyard 263

INDEX

Rousay haunting 14
Roy, Archie 128, 332–4
    'Glasgow's ghostbuster' 333
    mediums and clairvoyants 333
Royal Chapel of St Peter ad Vincula
      94, 145, 328
Royal Engineers 256
Royal Naval Volunteer
      Reserve 47
Royal Scots Fusiliers 242
Ruaidrí Ua Conchobair 20
Ruhr Valley 36–7, 38
Russell, Bertrand 45
Rydal Mount 256

sadness 337
Sagan, Carl 40
sailors 17
    hauntings of 11, 12
Saint-Germain-en-Laye 288
Saint-Ruhe, Lieutenant General
      Charles Chalmot de 23–5
Sampson, Agnes 57, 58
    ghost of 62
Sander, Nicholas 141
Sandwood Bay 9–17
    Am Buachaille 9
    corpses of mariners 12
    father and son stories 13, 15, 16,
      17, 341
    fishing boats tragedy 15
    hauntings
      bearded man 13, 14
      Gunn's mermaid 11
      sailors 11
      towering man 14–15

homesteads 10
shipwrecks 11
stories 15–16
Sandwood Cottage 9
    hauntings
      Australian visitor 11–12
      bearded sailor 12
      skittish horse 12
      thrown crockery 12
      towering man 14–15
sans-culottes 288–9
Sark, Battle of 153
Sarsfield, Major General Patrick 22,
      23–4, 25
'The Sash' 29
Savani et al. 40
Saxons 193
Scarp, island of 261, 262
Scots College 288
Scots Magazine 69
Scott, Captain Robert 246
Scott, James, 1st Duke of
      Monmouth 149
Scott, Sir Walter 102, 110–11, 196
Scruton, Roger 40–1
Seaforth, Earl of 138
Sebastopol 89–90
Sebastopol Bell 89–90
seers 128, 271
    Brahan Seer 129, 134, 135, 138, 264
    George Orwell 274, 276
A Sense of Something Strange
      (Roy) 333
Seven Men of Noidart 132
Seventh Day Adventism 114
Seymour, Jane 92, 93, 94, 145

INDEX

Sgurr Thormaid (Norman's Peak) 68

Shackleton, Ernest 239, 242, 245–7, 248–9

  Antarctica expedition 245–7, 248–9

  leadership 246

Shakespeare, William 96, 253, 277

Shap 255

Shaw, George Bernard 175

Shelley, Mary 234, 235

  ghost storytelling 235–6

Shelley, Percy Bysshe 234

Shepherd, Nan 66, 76

Sherriff, R. C. 32

Sherwood Forest 229

Shiloh 207

shipwrecks 11, 219

Shira, Indre 112, 117

Shirley (Brontë) 234

A Short History of the Tower of London (Younghusband) 141

Sidmouth, Lord 231

'Sign of the Times' 233–4

Simpson, Mr 35

Sir Gawain and the Green Knight 214–15

617 Squadron 38, 39

Skara Brae 317–18

skilled workers 228

Skye, Isle of 267, 269

  hauntings

    little black car 270–2

    roads 270

Slippery Stones 36

Smith, Mabel 170, 171

Smith, Reverend Guy Eric 170, 171

Society for Psychical Research 171, 200

Society of Friends 53, 61

Somerled 279, 283

Sorpe dam 37, 38

Sound of Scarp 261

South Georgia 239, 245, 246, 249

South (Shackleton) 239, 248–9

South Wales Ghost Stories 258–9

Southern Ocean 246

Spanish Armada 11

Spanish Civil War 274

Spilsbury, Bernard 175

spirits *see* ghosts

spiritualism 113–15, 126, 174

Squadron X 37, 38

St Albans, Suzanne 210

St Andrew's Church 257

St Columba (Colum Cille) 80–3, 83–4, 86–7, 88

  see also Iona

St James 35

St John 35

St John's hill 156

St Mahew's Chapel 262–3

St Marcellus's Flood 317

St Nectan's Glen 221

St Peter's Church 224–5, 226

St Philip's Church 36

Stag Boy (Rayner) 179

Starkey, David 93

stately homes 113, 115

  ghosts 115

  see also Raynham Hall

Stevenson, Robert 12

Stirling Castle 132

Stockholm Museum 64

stocking frames 228, 229

INDEX

Stone Age farmers 192
Stoneymollan coffin road 262–3
Stopes-Roe, Mary 39
storms 317
The Story of the British Isles in 100 Places (Oliver) 316
Strachey, Lytton and James 45
Strange, Bernard Le 124
String Road 263
String Theory 75–6, 78
Stuart, Charles Edward *see* Bonnie Prince Charlie
suffragettes 305
Sullivan, Danny 208–9
Sunday Express 33
Sunday Times 202
survivors' guilt 251
Swaledale Corpse Way 257
Swifte, Edmund Lenthal 147–8
Swindale Head 255
Swythamley 219
Swythamley and its Neighbourhood, Past and Present (Brocklehurst) 217
Synge, John Millington 266

Talbot, Richard, Earl of Tyrconnell 21, 22, 23
Talmud 201
Tandy, Vic 73, 74–5
    apparition 73
    low-frequency sound wave 73–4
Tantallon Castle 103
Te Kuri a Pawa (Pawa's dog) 187, 188
Teague Land: or a Merry Ramble to the Wild Irish (Dunton) 27

technology, fear of 236–7
Tectonic Strain Theory 221
Terra Nova expedition 246
Tewnion, Alexander 69–70, 71
The Road 263
There Is No Death (Marryat) 121
Theseus 296
thin places 128
    Celtic Christianity 40, 77, 78
    Iona 79–80, 83
The Third Man Factor: Surviving the Impossible (Geiger) 249, 251
third man phenomenon 249, 340
Thomas, Charles 86
Thomas, Dylan 15
Thomas, Mary 259
Three Nuns 226–7
    hauntings of Luddites 227
Threlkeld, Caleb 18–19, 26
Tigguo Cobauc 292, 294–6
timeslips 39
Tír na nÓg 262
Tiresias 296
Todd, Richard 39
tombs, ancient 191–2, 193
Topcliffe, Cecilia 227
Toronto City News 251
Torr an Aba (Hill of the Abbot) 86–7
Tower of London 93–4, 145–6
    burials 149–50
    hauntings
        Anne Boleyn 150
        Edward V 146
        glass tube apparition 147–8
        Guildford Dudley 146
        huge bear 148

Tower – *cont.*
  Lady Jane Grey 146
  Margaret Pole 146–7
  Richard, Duke of York 146
  whitish, female form 148
  Inner Ward 148–9
  restoration work 149
  royal chapel 149
The Town Below the Ground
  (Henderson) 155
Townshend, Charles, 2nd Viscount
  116, 117, 119–20
Townshend, Gwladys, Dowager
  Marchioness 123–4
Townshend, Horatio, 1st
  Viscount 116
Townshend, John 115–16
Townshend, John, 6th Marquess 124
Townshend, Lady Dorothy 'Dolly'
  119–20
Townshend, Sir Roger, 1st Baronet
  115, 116
transhumanism 237
tribes 79
Trilling, Lionel 275
True Ghost Stories (Townshend)
  123–4
Truss, Liz 309
Tuc d'Audoubert 297, 298
  bison carvings 298–9
Tudor, Margaret 103
tunnels 296, 297
  see also caves
The Turn of the Screw (James) 315, 318
Twain, Mark 40
Twogood, Chas 187–8

Uffington White Horse 193, 197, 198
Uí Niall (O'Neill) clan 80
Ukraine-Russian war 237
Under Milk Wood (Thomas) 15
*Underground: A Human History of
  the Worlds Beneath Our Feet*
  (Hunter) 299
Underwood, Peter 12, 13
underworld 296
Unknown Warrior 51
Unreliable Memoirs (James) 41
Upper Derwent valley 33, 35–6
Upper Glenn Luss 263

Vale House 322
Valley Forge 207
Vampyre (Polidori) 236
Vedbaek 64
Vernon, John 308
Victoria, Queen 90–1, 164
Vikings 10, 84
Villiers, George, Earl of Buckingham
  98–9
Vincent, Jack 245, 246
Votadini (later Gododdin) 152–3

Wainwright, Alfred 255
Wallis, Barnes Neville 37, 38
Walpole, Horace 120, 236
Walpole, Robert Sn. 119
Walpole, Sir Robert 119
War of 1812 224, 225
War of the Austrian Succession
  130, 132
War Sonnets *see* "nineteen fourteen"
  (Brooke)

INDEX

Ware, Fabian 51
Warner Brothers 321–2
Washington, George 207
Washington Post 210
The Waste Land (Eliot) 249, 252, 341–2
Waterloo, Battle of 224, 225
Watson, Susan 272
Wayland (Wolund) 191, 195, 197
Wayland's Smithy 191, 192, 194–5, 196, 197, 198
   hauntings 197
We Faked the Ghosts of Borley Rectory (Mayerling) 173
Wehlin, Joakim 222
Weiner, Eric 40
Weir, Jean 'Grizel' 158–9
Weir, Major Thomas 158–9
Wentworth-Day, James 108
West Dart river 179
Westminster Abbey 288
whaling station, Stromness 246, 247
Whalley Abbey 54–5
Wharton, Lord Thomas 119
Whitby Abbey 329
White Gate Inn 226, 227
   hauntings of Luddites 227
Whittle, Anne 59
Widecombe Fair 181
Wild Hunt 177–9, 181, 182, 184, 185
   coffin roads 178, 182
   hauntings
      dead infant son 181–2
   hunted king 184
The Wild Hunt of Hagworthy 179
Wilkinson, Jemima 114

William IV, King 94
William of Orange 19, 21–2, 29, 131, 281, 282, 283
William the Conqueror 90
Williamite War 19, 21, 22, 29
Windsor Castle 328
   fire 91
   haunted nature of 93
   hauntings
      Anne Boleyn 95–6
      bells in the Curfew Tower 97
      boy in the Deanery 97
      Charles I 99
      Elizabeth I 98
      George III 97–8
      George Villiers 98
      Henry VIII 94–5
      Herne the Hunter 96
      horned figure 97
      suicide of guardsman 97
      wraith 97
   history 90, 91
   repair and maintenance 90–1
   Round Tower 89, 90
   Sebastopol Bell 89–90
   St George's Chapel 93, 94–5
Wise, Francis 196
Wistman's Wood 179–81, 183, 184, 185
   infant son story 181–2
witchcraft 55–8
   Jean Weir 158–9
   Lady Janet Douglas 104, 110
   natural remedies 58–9
   Pendle Hill 54, 55–7, 59–60, 61
   Scotland 57–8
   trials 59–60

INDEX

Withnail and I 329, 337–8
Wolfe, Humbert 28
Wood, Wendy 69, 71
Woodchester Park 208
Woolf, Virginia 45
Wordsworth, William 184–5,
    256–7
World Trade Center 249–51
   North Tower 250
   South Tower 250
Worsley, Frank 245, 246, 249
The Worst Journey in the World
      (Cherry-Garrard) 246–7

wreckers 16
Wycliffe, John 216–17, 218

Ye Olde Salutation Inn 293
Ye Olde Trip to Jerusalem 293–4
   ghost (Rosie) 294
Yeats, William Butler 45
Young Nick's Head 187
Younghusband, George 141
Ypres 206, 211
yurei 187

Zeus 190–1

ABOUT THE AUTHOR

Neil Oliver was born in Renfrew, Scotland. He studied archaeology at the University of Glasgow and worked as an archaeologist before training as a journalist. Since 2002 he has presented various TV series including *Coast* (in the UK and in the Antipodes), *A History of Ancient Britain*, *Vikings* and *Sacred Wonders of Britain*. He is the author of several non-fiction books, including the bestsellers *The Story of Britain in 100 Places*, *The Story of the World in 100 Moments* and *Wisdom of the Ancients*, and one novel. He lives in Stirling with his wife, three children and two Irish wolfhounds.